Short Courses

 Microorganisms, Fungi, and Plants

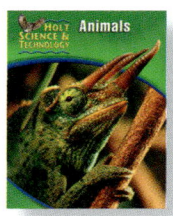 Animals

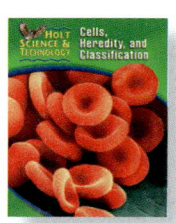 Cells, Heredity, and Classification

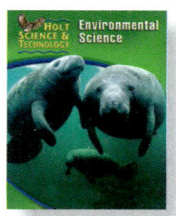 Environmental Science

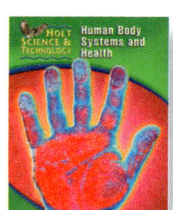 Human Body Systems and Health

 Inside the Restless Earth

 Earth's Changing Surface

 Water on Earth

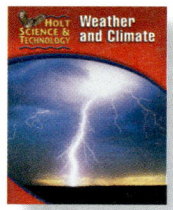 Weather and Climate

 Astronomy

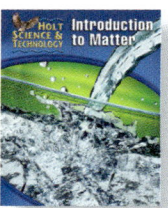 Introduction to Matter

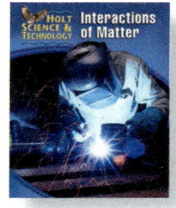 Interactions of Matter

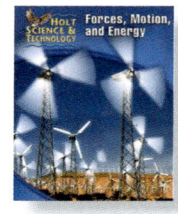 Forces, Motion, and Energy

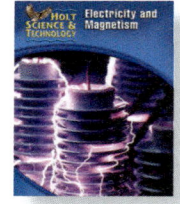 Electricity and Magnetism

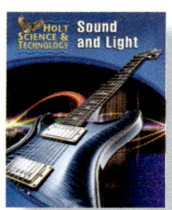 Sound and Light

Teacher Edition WALK-THROUGH

Student Edition CONTENTS IN BRIEF

HOLT, RINEHART AND WINSTON

A Harcourt Education Company

Orlando • **Austin** • New York • San Diego • Toronto • London

Designed to meet the needs of all students

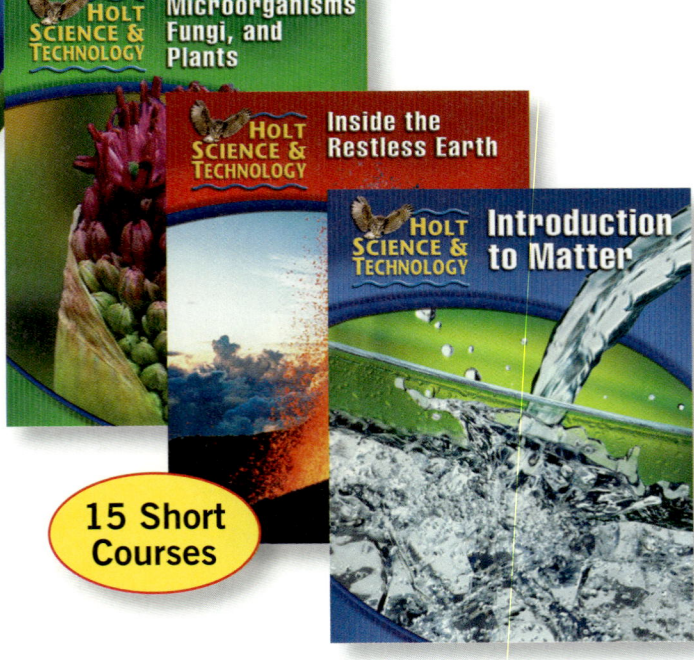

15 Short Courses

Holt Science & Technology: Short Course Series allows you to match your curriculum by choosing from 15 books covering life, earth, and physical sciences. The program reflects current curriculum developments and includes the strongest skills-development strand of any middle school science series. Students of all abilities will develop skills that they can use both in science as well as in other courses.

STUDENTS OF ALL ABILITIES RECEIVE THE READING HELP AND TAILORED INSTRUCTION THEY NEED.

- The *Student Edition* is accessible with a clean, easy-to-follow design and highlighted vocabulary words.
- Inclusion strategies and different learning styles help support all learners.
- Comprehensive **Section** and **Chapter Reviews** and **Standardized Test Preparation** allow students to practice their test-taking skills.
- **Reading Comprehension Guide** and **Guided Reading Audio CDs** help students better understand the content.

CROSS-DISCIPLINARY CONNECTIONS LET STUDENTS SEE HOW SCIENCE RELATES TO OTHER DISCIPLINES.

- **Mathematics, reading,** and **writing skills** are integrated throughout the program.
- Cross-discipline **Connection To** features show students how science relates to language arts, social studies, and other sciences.

A FLEXIBLE LABORATORY PROGRAM HELPS STUDENTS BUILD IMPORTANT INQUIRY AND CRITICAL-THINKING SKILLS.

- The laboratory program includes labs in each chapter, labs in the **LabBook** at the end of the text, six different lab books, and **Video Labs.**
- All labs are teacher-tested and rated by difficulty in the *Teacher Edition,* so you can be sure the labs will be appropriate for your students.
- A variety of labs, from **Inquiry Labs** to **Skills Practice Labs,** helps you meet the needs of your curriculum and work within the time constraints of your teaching schedule.

INTEGRATED TECHNOLOGY AND ONLINE RESOURCES EXPAND LEARNING BEYOND CLASSROOM WALLS.

- An **Enhanced Online Edition** or **CD-ROM Version** of the student text lightens your students' load.

- **SciLinks,** a Web service developed and maintained by the National Science Teachers Association (NSTA), contains current prescreened links directly related to the textbook.

- **Brain Food Video Quizzes** on videotape and DVD are game-show style quizzes that assess students' progress and motivate them to study.

- The **One-stop Planner® CD-ROM** with **Exam View® Test Generator** contains all of the resources you need including an *Interactive Teacher Edition,* worksheets, customizable lesson plans, **Holt Calendar Planner,** a powerful test generator, **Lab Materials QuickList Software,** and more.

- Spanish Resources include **Guided Reading Audio CD** in Spanish.

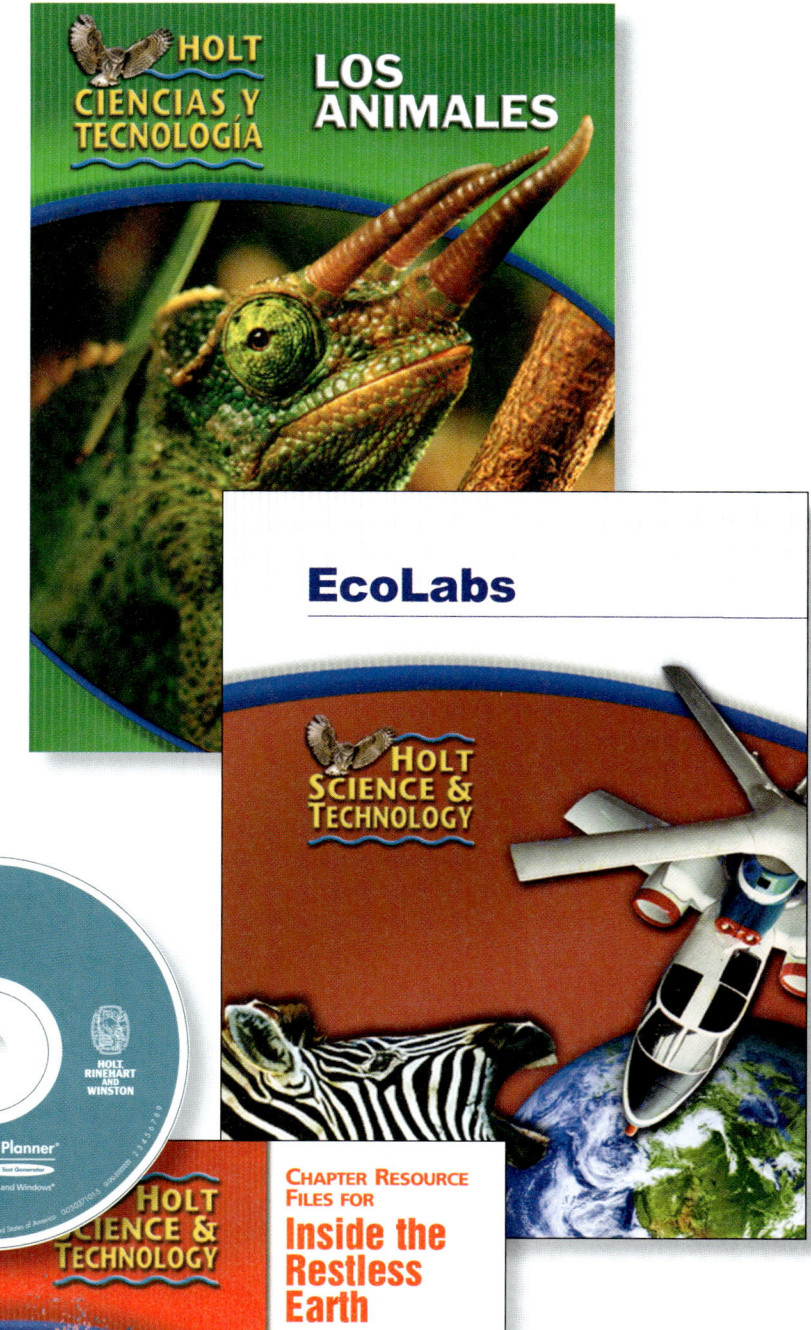

CHAPTER RESOURCE FILES FOR
Inside the Restless Earth

Skills Worksheets
- Directed Reading A
- Directed Reading B
- Vocabulary & Notes
- Section Reviews
- Chapter Review
- Reinforcement
- Critical Thinking

Assessments
- Section Quizzes
- Chapter Test A
- Chapter Test B
- Chapter Test C
- Performance-Based Assessment
- Standardized Test Preparation

Labs and Activities
- Datasheets for In-Text Labs
- Datasheets for Quick Labs
- Datasheets for LabBook
- Vocabulary Activity
- SciLinks® Activity

Teacher Resources
- Teacher Notes for Performance-Based Assessment
- Lab Notes and Answers
- Answer Keys
- Lesson Plans
- Test Item Listing for ExamView® Test Generator
- Teaching Transparencies
- Chapter Starter Transparencies
- Bellringer Transparencies
- Concept Mapping Transparencies

Life Science

PROGRAM SCOPE AND SEQUENCE

Selecting the right books for your course is easy. Just review the topics presented in each book to determine the best match to your district curriculum.

C

Cells: The Basic Units of Life
- Cells, tissues, and organs
- Populations, communities, and ecosystems
- Cell theory
- Surface-to-volume ratio
- Prokaryotic versus eukaryotic cells
- Cell organelles

The Cell in Action
- Diffusion and osmosis
- Passive versus active transport
- Endocytosis versus exocytosis
- Photosynthesis
- Cellular respiration and fermentation
- Cell cycle

Heredity
- Dominant versus recessive traits
- Genes and alleles
- Genotype, phenotype, the Punnett square and probability
- Meiosis
- Determination of sex

Genes and Gene Technology
- Structure of DNA
- Protein synthesis
- Mutations
- Heredity disorders and genetic counseling

The Evolution of Living Things
- Adaptations and species
- Evidence for evolution
- Darwin's work and natural selection
- Formation of new species

The History of Life on Earth
- Geologic time scale and extinctions
- Plate tectonics
- Human evolution

Classification
- Levels of classification
- Cladistic diagrams
- Dichotomous keys
- Characteristics of the six kingdoms

D

Body Organization and Structure
- Homeostasis
- Types of tissue
- Organ systems
- Structure and function of the skeletal system, muscular system, and integumentary system

Circulation and Respiration
- Structure and function of the cardiovascular system, lymphatic system, and respiratory system
- Respiratory disorders

The Digestive and Urinary Systems
- Structure and function of the digestive system
- Structure and function of the urinary system

Communication and Control
- Structure and function of the nervous system and endocrine system
- The senses
- Structure and function of the eye and ear

Reproduction and Development
- Asexual versus sexual reproduction
- Internal versus external fertilization
- Structure and function of the human male and female reproductive systems
- Fertilization, placental development, and embryo growth
- Stages of human life

Body Defenses and Disease
- Types of diseases
- Vaccines and immunity
- Structure and function of the immune system
- Autoimmune diseases, cancer, and AIDS

Staying Healthy
- Nutrition and reading food labels
- Alcohol and drug effects on the body
- Hygiene, exercise, and first aid

E

Interactions of Living Things
- Biotic versus abiotic parts of the environment
- Producers, consumers, and decomposers
- Food chains and food webs
- Factors limiting population growth
- Predator-prey relationships
- Symbiosis and coevolution

Cycles in Nature
- Water cycle
- Carbon cycle
- Nitrogen cycle
- Ecological succession

The Earth's Ecosystems
- Kinds of land and water biomes
- Marine ecosystems
- Freshwater ecosystems

Environmental Problems and Solutions
- Types of pollutants
- Types of resources
- Conservation practices
- Species protection

Energy Resources
- Types of resources
- Energy resources and pollution
- Alternative energy resources

Earth Science

F INSIDE THE RESTLESS EARTH	**G** EARTH'S CHANGING SURFACE
CHAPTER 1 **Minerals of the Earth's Crust** • Mineral composition and structure • Types of minerals • Mineral identification • Mineral formation and mining	**Maps as Models of the Earth** • Structure of a map • Cardinal directions • Latitude, longitude, and the equator • Magnetic declination and true north • Types of projections • Aerial photographs • Remote sensing • Topographic maps
CHAPTER 2 **Rocks: Mineral Mixtures** • Rock cycle and types of rocks • Rock classification • Characteristics of igneous, sedimentary, and metamorphic rocks	**Weathering and Soil Formation** • Types of weathering • Factors affecting the rate of weathering • Composition of soil • Soil conservation and erosion prevention
CHAPTER 3 **The Rock and Fossil Record** • Uniformitarianism versus catastrophism • Superposition • The geologic column and unconformities • Absolute dating and radiometric dating • Characteristics and types of fossils • Geologic time scale	**Agents of Erosion and Deposition** • Shoreline erosion and deposition • Wind erosion and deposition • Erosion and deposition by ice • Gravity's effect on erosion and deposition
CHAPTER 4 **Plate Tectonics** • Structure of the Earth • Continental drifts and sea floor spreading • Plate tectonics theory • Types of boundaries • Types of crust deformities	
CHAPTER 5 **Earthquakes** • Seismology • Features of earthquakes • P and S waves • Gap hypothesis • Earthquake safety	
CHAPTER 6 **Volcanoes** • Types of volcanoes and eruptions • Types of lava and pyroclastic material • Craters versus calderas • Sites and conditions for volcano formation • Predicting eruptions	

H WATER ON EARTH

The Flow of Fresh Water
- Water cycle
- River systems
- Stream erosion
- Life cycle of rivers
- Deposition
- Aquifers, springs, and wells
- Ground water
- Water treatment and pollution

Exploring the Oceans
- Properties and characteristics of the oceans
- Features of the ocean floor
- Ocean ecology
- Ocean resources and pollution

The Movement of Ocean Water
- Types of currents
- Characteristics of waves
- Types of ocean waves
- Tides

I WEATHER AND CLIMATE

The Atmosphere
- Structure of the atmosphere
- Air pressure
- Radiation, convection, and conduction
- Greenhouse effect and global warming
- Characteristics of winds
- Types of winds
- Air pollution

Understanding Weather
- Water cycle
- Humidity
- Types of clouds
- Types of precipitation
- Air masses and fronts
- Storms, tornadoes, and hurricanes
- Weather forecasting
- Weather maps

Climate
- Weather versus climate
- Seasons and latitude
- Prevailing winds
- Earth's biomes
- Earth's climate zones
- Ice ages
- Global warming
- Greenhouse effect

J ASTRONOMY

Studying Space
- Astronomy
- Keeping time
- Types of telescope
- Radioastronomy
- Mapping the stars
- Scales of the universe

Stars, Galaxies, and the Universe
- Composition of stars
- Classification of stars
- Star brightness, distance, and motions
- H-R diagram
- Life cycle of stars
- Types of galaxies
- Theories on the formation of the universe

Formation of the Solar System
- Birth of the solar system
- Structure of the sun
- Fusion
- Earth's structure and atmosphere
- Planetary motion
- Newton's Law of Universal Gravitation

A Family of Planets
- Properties and characteristics of the planets
- Properties and characteristics of moons
- Comets, asteroids, and meteoroids

Exploring Space
- Rocketry and artificial satellites
- Types of Earth orbit
- Space probes and space exploration

Physical Science

K INTRODUCTION TO MATTER	**L** INTERACTIONS OF MATTER
CHAPTER 1	
The Properties of Matter • Definition of matter • Mass and weight • Physical and chemical properties • Physical and chemical change • Density	**Chemical Bonding** • Types of chemical bonds • Valence electrons • Ions versus molecules • Crystal lattice
CHAPTER 2	
States of Matter • States of matter and their properties • Boyle's and Charles's laws • Changes of state	**Chemical Reactions** • Writing chemical formulas and equations • Law of conservation of mass • Types of reactions • Endothermic versus exothermic reactions • Law of conservation of energy • Activation energy • Catalysts and inhibitors
CHAPTER 3	
Elements, Compounds, and Mixtures • Elements and compounds • Metals, nonmetals, and metalloids (semiconductors) • Properties of mixtures • Properties of solutions, suspensions, and colloids	**Chemical Compounds** • Ionic versus covalent compounds • Acids, bases, and salts • pH • Organic compounds • Biomolecules
CHAPTER 4	
Introduction to Atoms • Atomic theory • Atomic model and structure • Isotopes • Atomic mass and mass number	**Atomic Energy** • Properties of radioactive substances • Types of decay • Half-life • Fission, fusion, and chain reactions
CHAPTER 5	
The Periodic Table • Structure of the periodic table • Periodic law • Properties of alkali metals, alkaline-earth metals, halogens, and noble gases	
CHAPTER 6	

M FORCES, MOTION, AND ENERGY

Matter in Motion
- Speed, velocity, and acceleration
- Measuring force
- Friction
- Mass versus weight

Forces in Motion
- Terminal velocity and free fall
- Projectile motion
- Inertia
- Momentum

Forces in Fluids
- Properties in fluids
- Atmospheric pressure
- Density
- Pascal's principle
- Buoyant force
- Archimedes' principle
- Bernoulli's principle

Work and Machines
- Measuring work
- Measuring power
- Types of machines
- Mechanical advantage
- Mechanical efficiency

Energy and Energy Resources
- Forms of energy
- Energy conversions
- Law of conservation of energy
- Energy resources

Heat and Heat Technology
- Heat versus temperature
- Thermal expansion
- Absolute zero
- Conduction, convection, radiation
- Conductors versus insulators
- Specific heat capacity
- Changes of state
- Heat engines
- Thermal pollution

N ELECTRICITY AND MAGNETISM

Introduction to Electricity
- Law of electric charges
- Conduction versus induction
- Static electricity
- Potential difference
- Cells, batteries, and photocells
- Thermocouples
- Voltage, current, and resistance
- Electric power
- Types of circuits

Electromagnetism
- Properties of magnets
- Magnetic force
- Electromagnetism
- Solenoids and electric motors
- Electromagnetic induction
- Generators and transformers

Electronic Technology
- Properties of semiconductors
- Integrated circuits
- Diodes and transistors
- Analog versus digital signals
- Microprocessors
- Features of computers

O SOUND AND LIGHT

The Energy of Waves
- Properties of waves
- Types of waves
- Reflection and refraction
- Diffraction and interference
- Standing waves and resonance

The Nature of Sound
- Properties of sound waves
- Structure of the human ear
- Pitch and the Doppler effect
- Infrasonic versus ultrasonic sound
- Sound reflection and echolocation
- Sound barrier
- Interference, resonance, diffraction, and standing waves
- Sound quality of instruments

The Nature of Light
- Electromagnetic waves
- Electromagnetic spectrum
- Law of reflection
- Absorption and scattering
- Reflection and refraction
- Diffraction and interference

Light and Our World
- Luminosity
- Types of lighting
- Types of mirrors and lenses
- Focal point
- Structure of the human eye
- Lasers and holograms

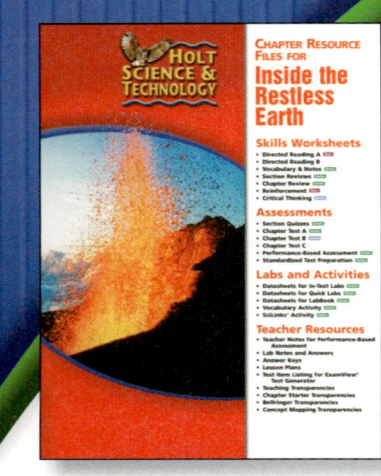

Program resources make teaching and learning easier.

CHAPTER RESOURCES

A *Chapter Resources book* accompanies each of the 15 *Short Courses*. Here you'll find everything you need to make sure your students are getting the most out of learning science—all in one book.

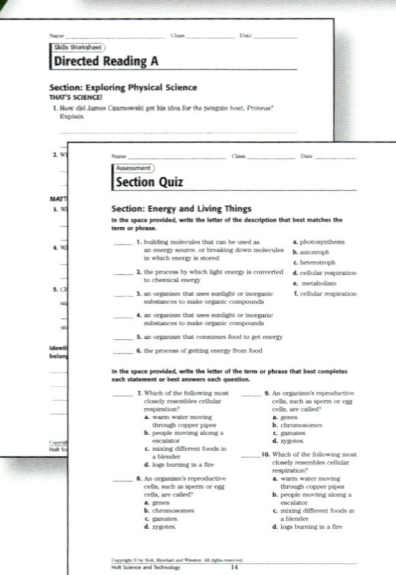

Skills Worksheets

- Directed Reading A: Basic
- Directed Reading B: Special Needs
- Vocabulary and Chapter Summary
- Section Reviews
- Chapter Reviews
- Reinforcement
- Critical Thinking

Labs & Activities

- Datasheets for Chapter Labs
- Datasheets for Quick Labs
- Datasheets for LabBook
- Vocabulary Activity
- SciLinks® Activity

Assessments

- Section Quizzes
- Chapter Tests A: General
- Chapter Tests B: Advanced
- Chapter Tests C: Special Needs
- Performance-Based Assessments
- Standardized Test Preparation

Teacher Resources

- Lab Notes and Answers
- Teacher Notes for Performance-Based Assessment
- Answer Keys
- Lesson Plans
- Test Item Listing for ExamView® Test Generator
- Full-color **Teaching Transparencies,** plus section **Bellringers, Concept Mapping,** and **Chapter Starter Transparencies.**

SPANISH RESOURCES

Spanish materials are available for each *Short Course:*

- *Student Edition*
- *Spanish Resources* booklet contains worksheets and assessments translated into Spanish with an English **Answer Key.**
- **Guided Reading Audio CD Program**

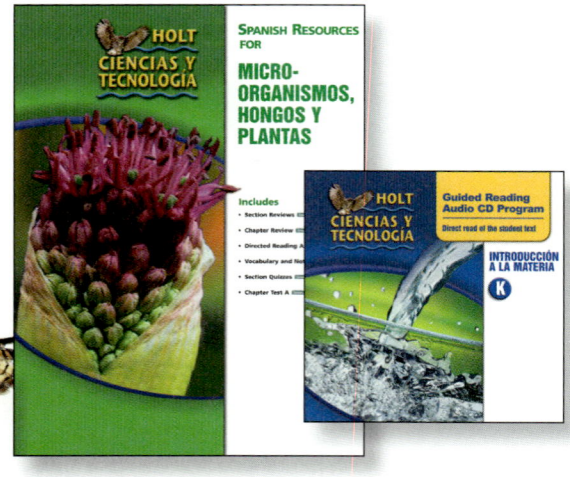

ONLINE RESOURCES

- *Enhanced Online Editions* engage students and assist teachers with a host of interactive features that are available anytime and anywhere you can connect to the Internet.
- **CNNStudentNews.com** provides award-winning news and information for both teachers and students.
- **SciLinks**—a Web service developed and maintained by the National Science Teachers Association—links you and your students to up-to-date online resources directly related to chapter topics.
- **go.hrw.com** links you and your students to online chapter activities and resources.
- **Current Science** articles relate to students' lives.

ADDITIONAL LAB AND SKILLS RESOURCES

- *Calculator-Based Labs* incorporates scientific instruments, offering students insight into modern scientific investigation.
- *EcoLabs & Field Activities* develops awareness of the natural world.
- *Holt Science Skills Workshop: Reading in the Content Area* contains exercises that target reading skills key.
- *Inquiry Labs* taps students' natural curiosity and creativity with a focus on the process of discovery.
- *Labs You Can Eat* safely incorporates edible items into the classroom.
- *Long-Term Projects & Research Ideas* extends and enriches lessons.
- *Math Skills for Science* provides additional explanations, examples, and math problems so students can develop their skills.
- *Science Skills Worksheets* helps your students hone important learning skills.
- *Whiz-Bang Demonstrations* gets your students' attention at the beginning of a lesson.

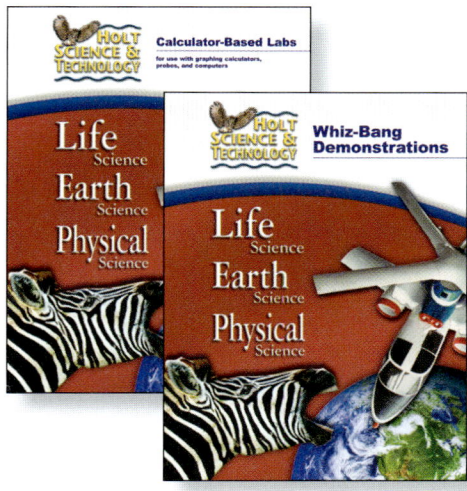

ADDITIONAL RESOURCES

- *Assessment Checklists & Rubrics* gives you guidelines for evaluating students' progress.
- *Holt Anthology of Science Fiction* sparks your students' imaginations with thought-provoking stories.
- *Holt Science Posters* visually reinforces scientific concepts and themes with seven colorful posters including **The Periodic Table of the Elements.**

- *Professional Reference for Teachers* contains professional articles that discuss a variety of topics, such as classroom management.
- *Program Introduction Resource File* explains the program and its features and provides several additional references, including lab safety, scoring rubrics, and more.
- *Science Fair Guide* gives teachers, students, and parents tips for planning and assisting in a science fair.
- *Science Puzzlers, Twisters & Teasers* activities challenge students to think about science concepts in different ways.

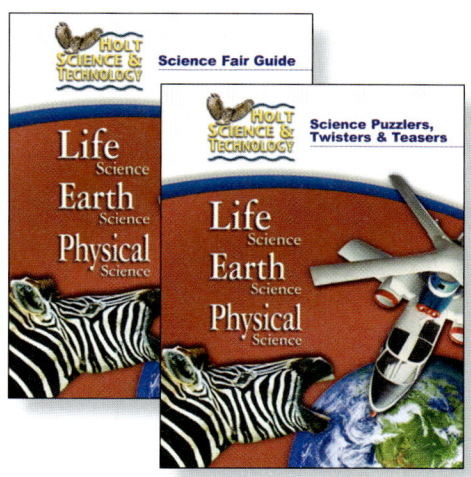

TECHNOLOGY RESOURCES

- *CNN Presents Science in the News: Video Library* helps students see the impact of science on their everyday lives with actual news video clips.
  - Multicultural Connections
  - Science, Technology & Society
  - Scientists in Action
  - Eye on the Environment
- *Guided Reading Audio CD Program*, available in English and Spanish, provides students with a direct read of each section.
- *HRW Earth Science Videotape* takes your students on a geology "field trip" with full-motion video.
- *Interactive Explorations CD-ROM Program* develops students' inquiry and decision-making skills as they investigate science phenomena in a virtual lab setting.

- *One-Stop Planner CD-ROM*® organizes everything you need on one disc, including printable worksheets, customizable lesson plans, a powerful test generator, **PowerPoint**® **LectureNotes, Lab Materials QuickList Software, Holt Calendar Planner, Interactive Teacher Edition,** and more.
- *Science Tutor CD-ROMs* help students practice what they learn with immediate feedback.
- *Lab Videos* make it easier to integrate more experiments into your lessons without the preparation time and costs. Available on DVD and VHS.
- **Brain Food Video Quizzes** are game-show style quizzes that assess students' progress. Available on DVD and VHS.
- *Visual Concepts CD-ROMs* include graphics, animations, and movie clips that demonstrate key chapter concepts.

Science and Math Worksheets

The **Holt Science & Technology** program helps you meet the needs of a wide variety of students, regardless of their skill level. The following pages provide examples of the worksheets available to improve your students' science and math skills whether they already have a strong science and math background or are weak in these areas. Samples of assessment checklists and rubrics are also provided.

In addition to the skills worksheets represented here, **Holt Science & Technology** provides a variety of worksheets that are correlated directly with each chapter of the program. Representations of these worksheets are found at the beginning of each chapter in this *Teacher Edition.*

Many worksheets are also available on the Holt Web site. The address is **go.hrw.com.**

Science Skills Worksheets: Thinking Skills

BEING FLEXIBLE

USING YOUR SENSES
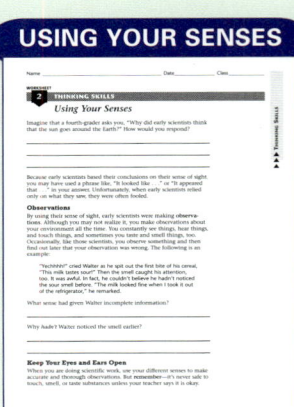

THINKING OBJECTIVELY
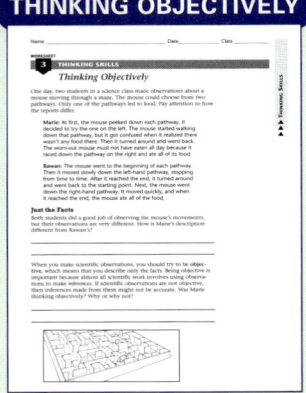

UNDERSTANDING BIAS
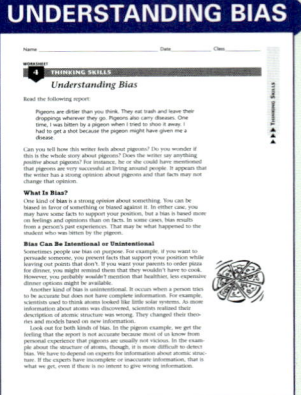

USING LOGIC
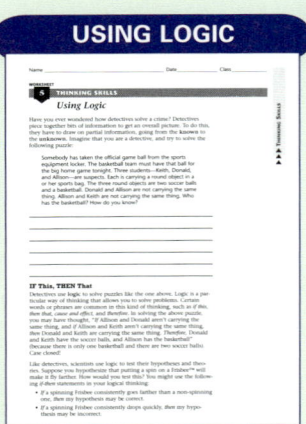

BOOSTING YOUR MEMORY
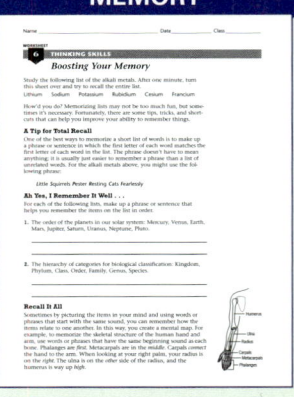

IMPROVING YOUR STUDY HABITS
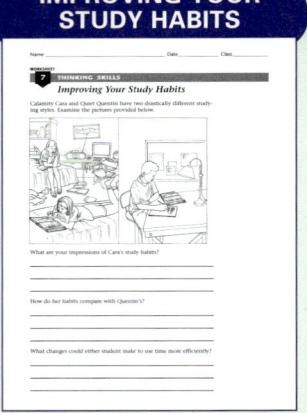

READING A SCIENCE TEXTBOOK

Science Skills Worksheets: Experimenting Skills

SAFETY RULES!
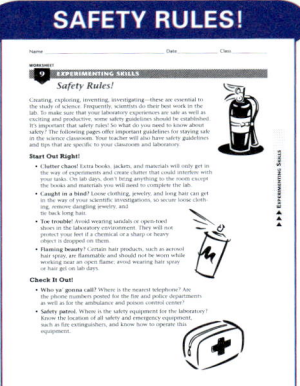

DOING A LAB WRITE-UP
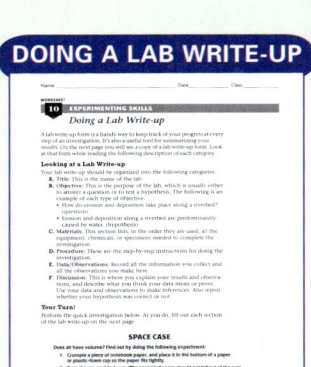

UNDERSTANDING VARIABLES
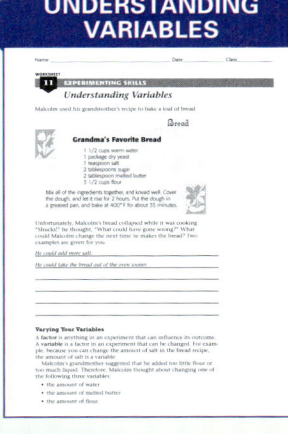

WORKING WITH HYPOTHESES
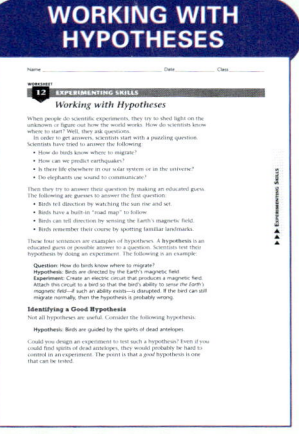

DESIGNING AN EXPERIMENT
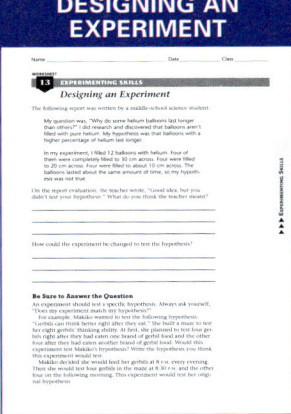

USING THE INTERNATIONAL SYSTEM OF UNITS (SI)
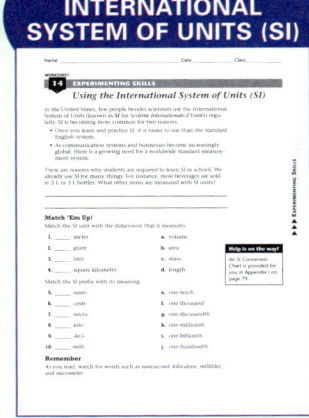

MEASURING
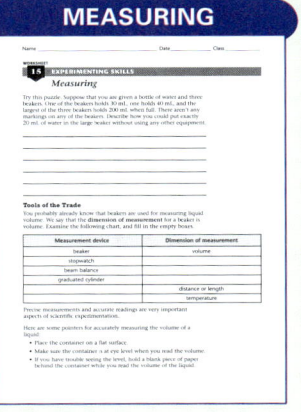

Science Skills Worksheets: Researching Skills

CHOOSING YOUR TOPIC
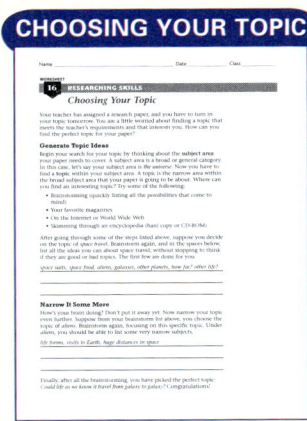

ORGANIZING YOUR RESEARCH
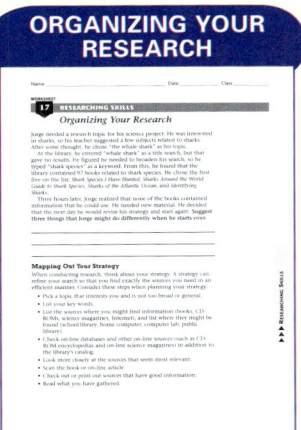

FINDING USEFUL SOURCES
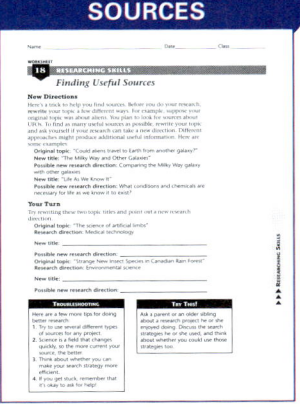

RESEARCHING ON THE WEB

Science Skills Worksheets: Researching Skills (continued)

IDENTIFYING BIAS

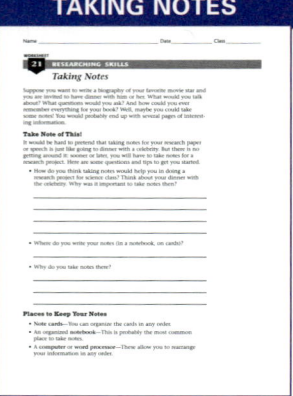

TAKING NOTES

Science Skills Worksheets: Communicating Skills

SCIENCE WRITING

SCIENCE DRAWING

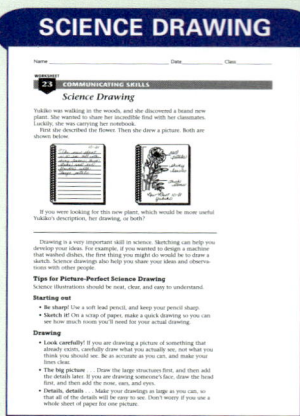

USING MODELS TO COMMUNICATE

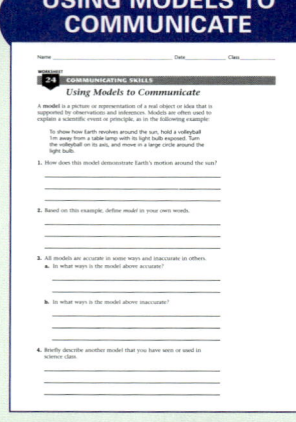

INTRODUCTION TO GRAPHS

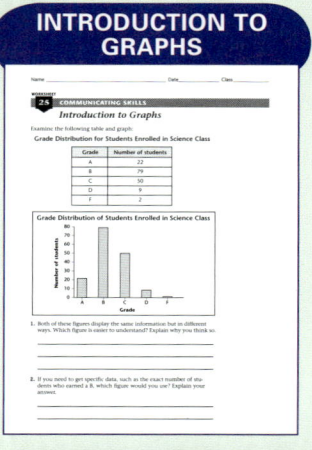

GRASPING GRAPHING

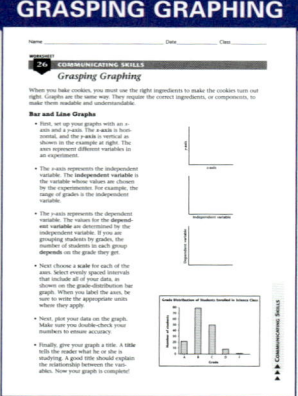

INTERPRETING YOUR DATA

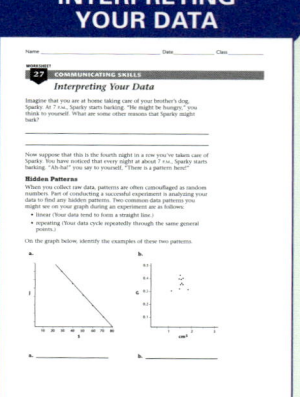

RECOGNIZING BIAS IN GRAPHS

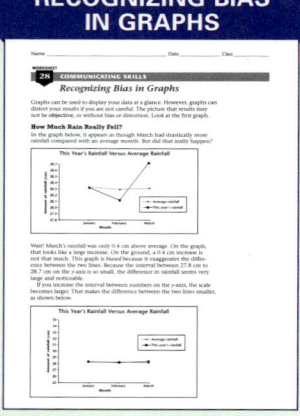

MAKING DATA MEANINGFUL

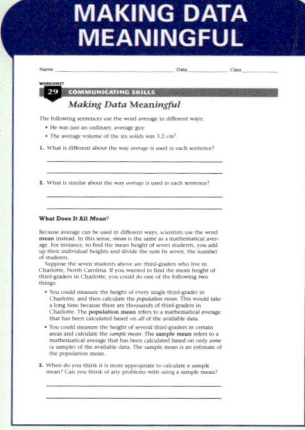

HINTS FOR ORAL PRESENTATIONS

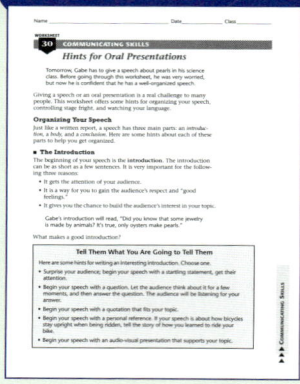

Math Skills for Science

ADDITION AND SUBTRACTION

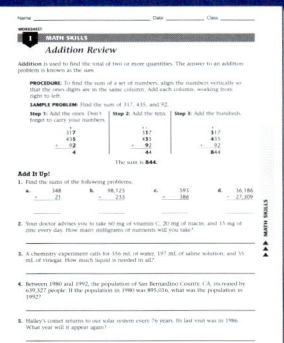

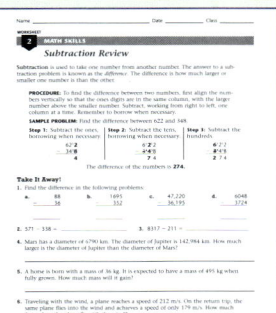

MULTIPLICATION

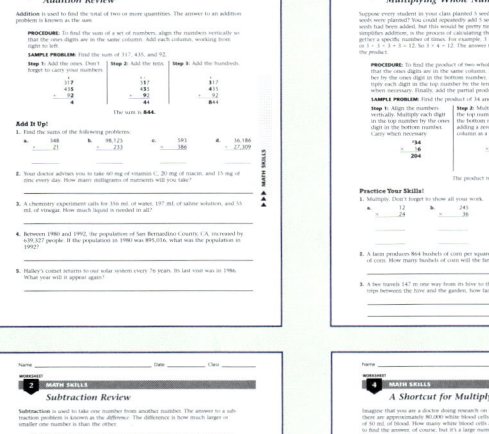
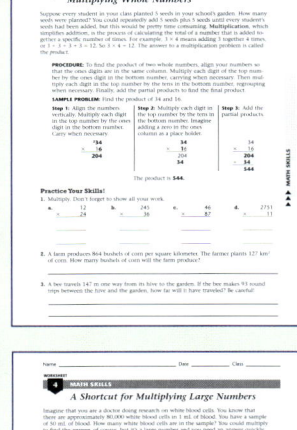

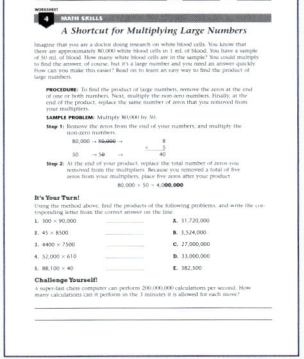

DIVISION

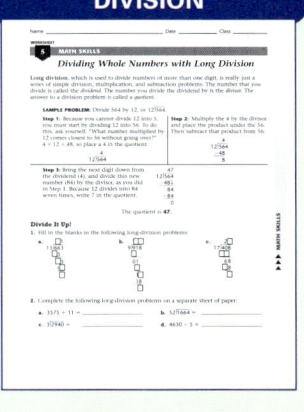

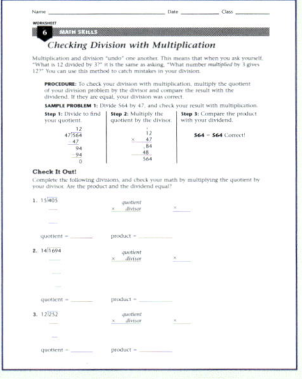

AVERAGES

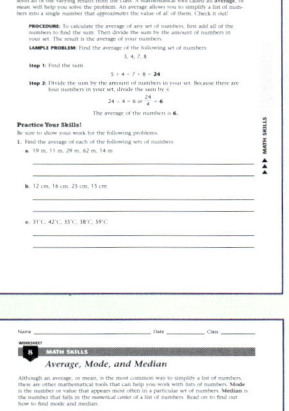

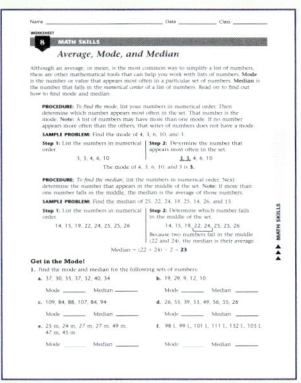

POSITIVE AND NEGATIVE NUMBERS

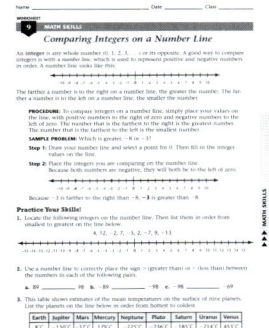

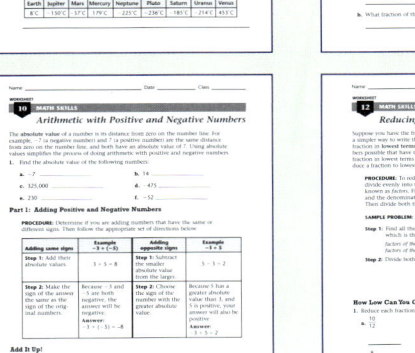

FRACTIONS

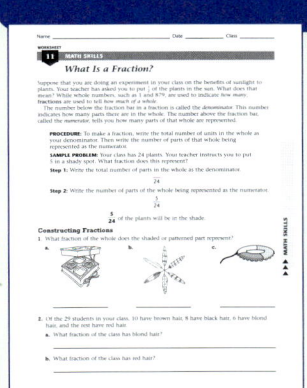

Math Skills for Science (continued)

RATIOS AND PROPORTIONS

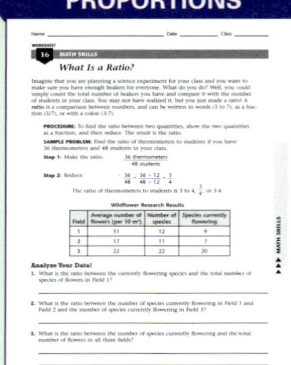

What Is a Ratio?

Using Proportions and Cross-Multiplication

DECIMALS

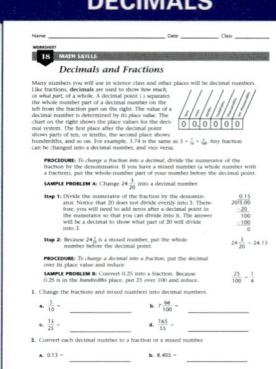

Decimals and Fractions

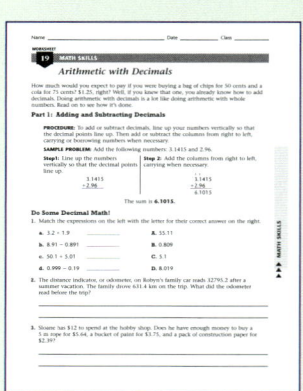

Arithmetic with Decimals

PERCENTAGES

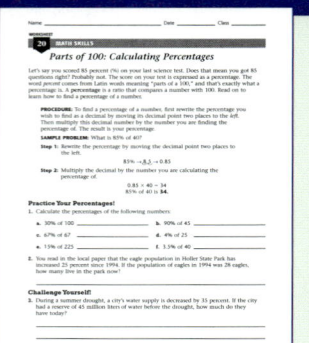

Parts of 100: Calculating Percentages

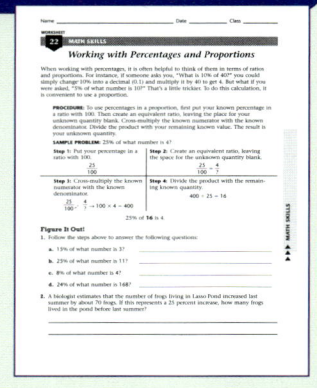

Working with Percentages and Proportions

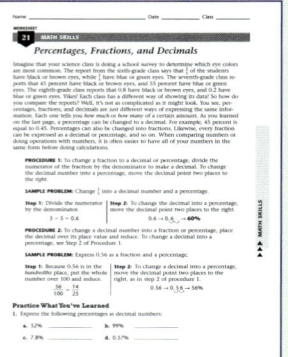

Percentages, Fractions, and Decimals

POWERS OF 10

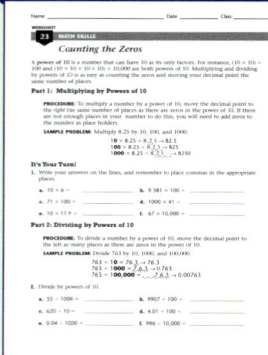

Counting the Zeros

Creating Exponents

SCIENTIFIC NOTATION

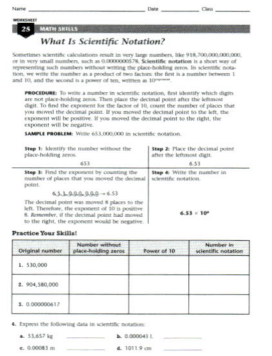

What Is Scientific Notation?

Multiplying and Dividing in Scientific Notation

SI MEASUREMENT AND CONVERSION

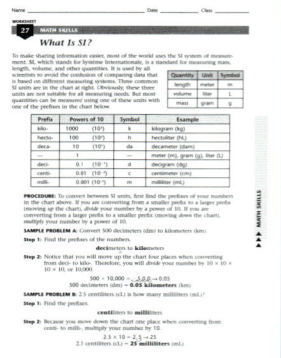

What Is SI?

A Formula for SI Catch-up

Math Skills for Science (continued)

GEOMETRY

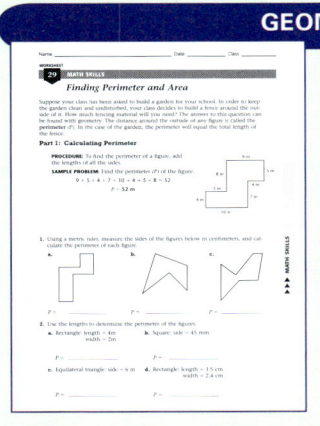 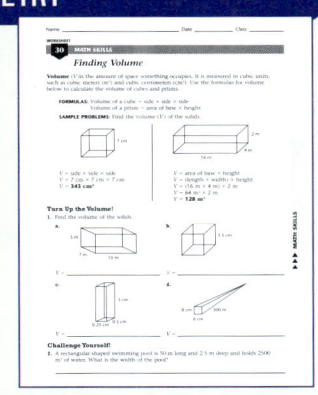

THE UNIT FACTOR AND DIMENSIONAL ANALYSIS

MATH IN SCIENCE: INTEGRATED SCIENCE

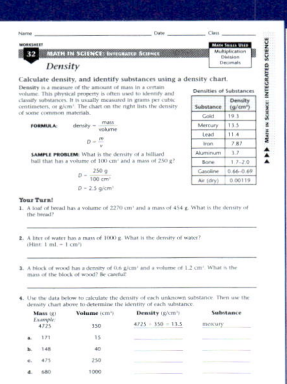 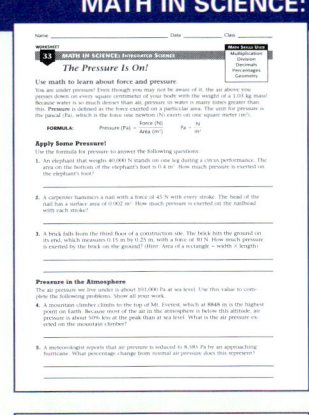 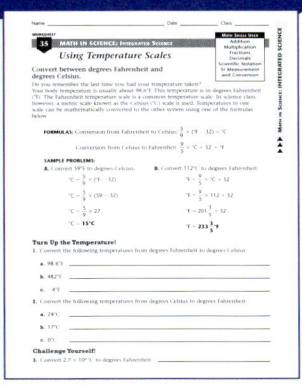

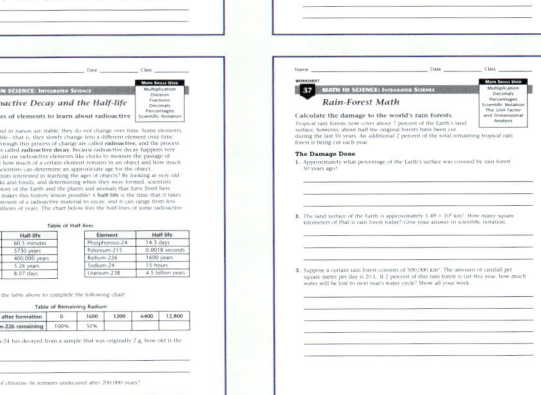

Math Skills for Science (continued)

MATH IN SCIENCE: LIFE SCIENCE

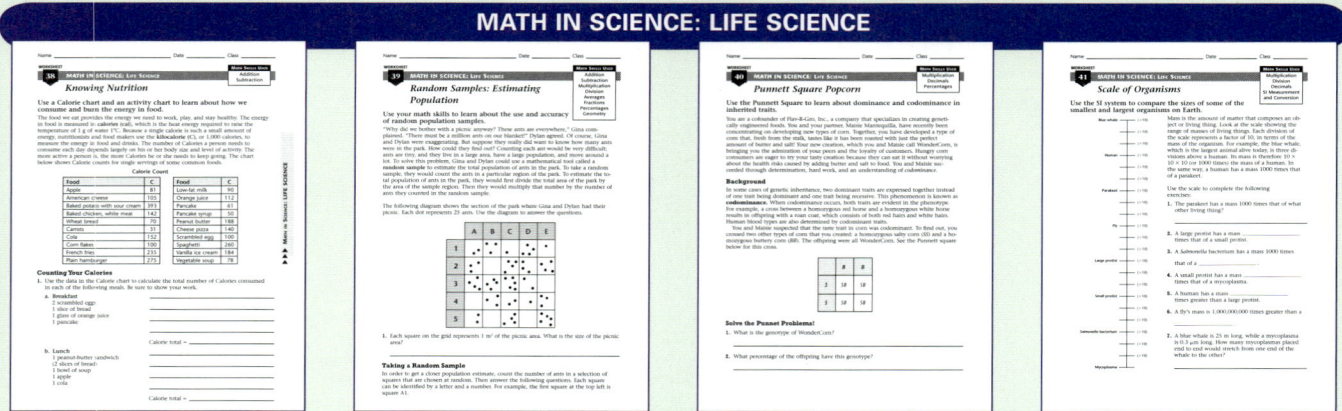

MATH IN SCIENCE: EARTH SCIENCE

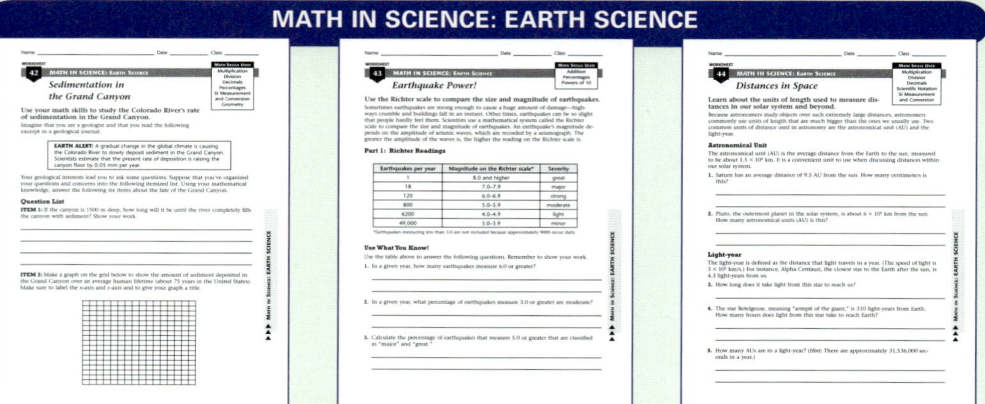

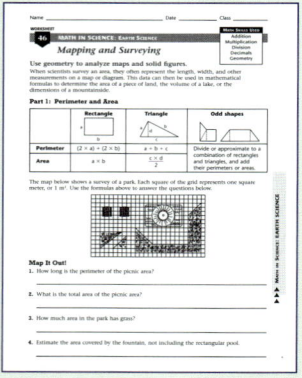

T18

Math Skills for Science (continued)

MATH IN SCIENCE: PHYSICAL SCIENCE

Worksheet 47 — Average Speed in a Pinewood Derby

Worksheet 48 — Newton: Force and Motion

Worksheet 49 — Momentum

Worksheet 50 — Balancing Chemical Equations

Worksheet 51 — Work and Power

Worksheet 52 — A Bicycle Trip

Worksheet 53 — Mechanical Advantage

Worksheet 54 — Color at Light Speed

Assessment Checklist & Rubrics

The following is just a sample of over 50 checklists and rubrics contained in this booklet.

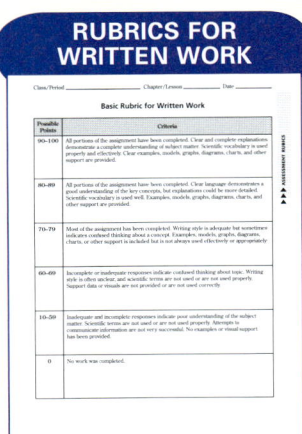

RUBRICS FOR WRITTEN WORK

Basic Rubric for Written Work

RUBRIC FOR EXPERIMENTS

Rubric for Experiments

TEACHER EVALUATION OF COOPERATIVE LEARNING

Teacher Evaluation of Cooperative Group Activity

TEACHER EVALUATION OF STUDENT PROGRESS

Teacher Evaluation of Daily Progress: Homework and Class Participation

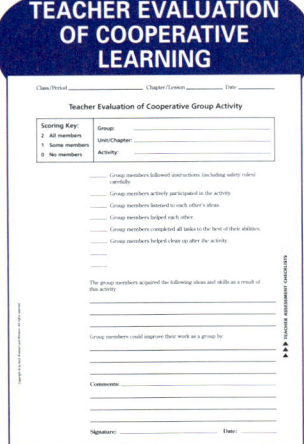

National Science Education Standards

The following lists show the chapter correlation of *Holt Science & Technology: Water on Earth* with the *National Science Education Standards* (grades 5–8).

Unifying Concepts and Processes

Standard	Chapter Correlation	
Systems, order, and organization Code: UCP 1	Chapter 1 Chapter 2 Chapter 3	1.1 2.1, 2.3 3.3
Evidence, models, and explanation Code: UCP 2	Chapter 1 Chapter 2 Chapter 3	1.1, 1.4 2.1, 2.2, 2.5 3.1, 3.2, 3.3, 3.4
Change, constancy, and measurement Code: UCP 3	Chapter 1 Chapter 2 Chapter 3	1.4 2.1, 2.2, 2.5 3.3
Form and function Code: UCP 5	Chapter 1	1.1

Science as Inquiry

Standard	Chapter Correlation	
Abilities necessary to do scientific inquiry Code: SAI 1	Chapter 1 Chapter 2 Chapter 3	1.4 2.2, 2.4, 2.5 3.1, 3.2, 3.3, 3.4
Understandings about scientific inquiry Code: SAI 2	Chapter 2	2.2

Science and Technology

Standard	Chapter Correlation	
Abilities of technological design Code: ST 1	Chapter 1 Chapter 3	1.1, 1.3 3.4
Understandings about science and technology Code: ST 2	Chapter 1 Chapter 2	1.4 2.2, 2.4, 2.5

Science in Personal Perspectives

Standard	Chapter Correlation	
Populations, resources, and environments Code: SPSP 2	Chapter 2	2.4, 2.5
Natural hazards Code: SPSP 3	Chapter 1 Chapter 2 Chapter 3	1.2, 1.2, 1.4 2.5 3.3
Risks and benefits Code: SPSP 4	Chapter 1 Chapter 2 Chapter 3	1.2 2.4, 2.5 3.3
Science and technology in society Code: SPSP 5	Chapter 1 Chapter 2	1.4 2.2, 2.4, 2.5

History and Nature of Science

Standard	Chapter Correlation	
Science as a human endeavor Code: HNS 1	Chapter 2 Chapter 3	2.2, 2.4, 2.5 3.1, 3.4
History of science Code: HNS 3	Chapter 2 Chapter 3	2.2 3.1, 3.4

Earth Science Content Standards

Structure of the Earth System

Standard	Chapter Correlation	
Lithospheric plates on the scales of continents and oceans constantly move at rates of centimeters per year in response to movements in the mantle. Major geological events, such as earthquakes, volcanic eruptions, and mountain building result from these plate motions. Code: ES 1b	**Chapter 2** **Chapter 3**	2.1, 2.2 3.3
Land forms are the result of a combination of constructive and destructive forces. Constructive forces include crustal deformation, volcanic eruption, and deposition of sediment, while destructive forces include weathering and erosion. Code: ES 1c	**Chapter 1**	1.1, 1.2, 1.3
Water, which covers the majority of the earth's surface, circulates through the crust, oceans, and atmosphere in what is known as the "water cycle." Water evaporates from the earth's surface, rises and cools as it moves to higher elevations, condenses as rain or snow, and falls to the surface where it collects in lakes, oceans soil, and in rocks underground. Code: ES 1f	**Chapter 1** **Chapter 2**	1.1 2.1
Water is a solvent. As it passes through the water cycle it dissolves minerals and gases and carries them to the oceans. Code: ES 1g	**Chapter 2**	2.1
The atmosphere is a mixture of nitrogen, oxygen, and trace gases that include water vapor. The atmosphere has different properties at different elevations. Code: ES 1h	**Chapter 2**	2.1
Global patterns of atmospheric movement influence local weather. Oceans have a major effect on climate, because water in the oceans holds a large amount of heat. Code: ES 1j	**Chapter 2** **Chapter 3**	2.1 3.1, 3.2

Earth's History

Standard	Chapter Correlation	
The earth processes we see today, including erosion, movement of lithospheric plates, and changes in atmospheric composition, are similar to those that occurred in the past. Earth history is also influenced by occasional catastrophes, such as the impact of an asteroid or comet. Code: ES 2a	**Chapter 2**	2.1

Earth in the Solar System

Standard	Chapter Correlation	
Gravity is the force that keeps planets in orbit around the sun and governs the rest of the motion in the solar system. Gravity alone holds us to the earth's surface and explains the phenomena of the tides. Code: ES 3c	**Chapter 3**	3.4

HOLT SCIENCE & TECHNOLOGY

Water on Earth

HOLT, RINEHART AND WINSTON

A Harcourt Education Company

Orlando • Austin • New York • San Diego • Toronto • London

Acknowledgments

Contributing Authors

Kathleen Kaska
ormer Lie and Earth Science Teacher and Science Department Chair

Robert J. Sager M.S. J.D. L.G.
Coordinator and roessor o Earth Science
Pierce College
Lakewood, Washington

nclusion Specialist

Karen Clay
nclsion Specialist Consltant
Boston, Massachusetts

Saety Reviewer

Jack Gerlovich Ph.D.
ssociate roessor
School of Education
Drake niversity
Des Moines, Iowa

Academic Reviewers

Kenneth H. Brink Ph.D.
Senior Scientist and hsical ceanoraph Director
Coastal Ocean Institute and Rinehart Coastal Research Center
Woods Hole Oceanographic Institution
Woods Hole, Massachusetts

John Brockhaus Ph.D.
roessor o eospatial normation Science and Director o eospatial normation Science ror am
Department of Geography and Environmental Engineering
nited States Military Academy
West Point, New York

Steven A. Jennings Ph.D.
ssociate roessor
Geography and Environmental Studies
niversity of Colorado at Colorado Springs
Colorado Springs, Colorado

Madeline Micceri Mignone Ph.D.
ssistant roessor
Natural Science
Dominican College
Orangeburg, New York

Kenneth H. Rubin P h.D.
ssociate roessor
Department of Geology Geophysics
niversity of Hawaii at Manoa
Honolulu, Hawaii

Teacher Reviewers

Diedre S. Adams
hsical Science nstrctor
Science Department
West Vigo Middle School
West Terre Haute, Indiana

ii

Robin K. Clanton
Science Department Head
Berrien Middle School
Nashville, Georgia

Laura Kitselman
*Science Teacher and
    Coordinator*
Loudoun Country Day
    School
Leesburg, Virginia

Sally M. Lesley
ESL Science Teacher
Burnet Middle School
Austin, Texas

Lab Development

Kenneth E. Creese
Science Teacher
White Mountain Junior
    High School
Rock Spring, Wyoming

Linda A. Culp
*Science Teacher and
    Department Chair*
Thorndale High School
Thorndale, Texas

Bruce M. Jones
*Science Teacher and
    Department Chair*
The Blake School
Minneapolis, Minnesota

Shannon Miller
Science and Math Teacher
Llano Junior High School
Llano, Texas

Robert Stephen Ricks
Special Services Teacher
Department of Classroom
    Improvement
Alabama State Department
    of Education
Montgomery, Alabama

James J. Secosky
Science Teacher
Bloomfield Central School
Bloomfield, New York

Lab Testing

Kenneth Creese
Science Teacher
White Mountain Junior
    High
Rock Springs, Wyoming

Norman Holcomb
Science Teacher
Marion Local Schools
Maria Stein, Ohio

Tracy Jahn
Science Teacher
Berkshire Jr–Sr. High
Canaan, New York

David M. Sparks
Science Teacher
Redwater Junior High
    School
Redwater, Texas

Gordon Zibelman
Science Teacher
Drexel Hill Middle School
Drexel Hill, Pennsylvania

Answer Checking

Catherine Podeszwa
Duluth, Minnesota

Feature Development

Katy Z. Allen
Hatim Belyamani
John A. Benner
David Bradford
Jennifer Childers
Mickey Coakley
Susan Feldkamp
Jane Gardner
Erik Hahn
Christopher Hess
Deena Kalai
Charlotte W. Luongo, MSc
Michael May
Persis Mehta, Ph.D.
Eileen Nehme, MPH
Catherine Podeszwa
Dennis Rathnaw
Daniel B. Sharp
John Stokes
April Smith West
Molly F. Wetterschneider

iii

H Water on Earth

Contents   **v**

Labs and Activities

How to Use Your Textbook

Your Roadmap for Success with Holt Science and Technology

Reading Warm-Up

A Reading Warm-Up at the beginning of every section provides you with the section's objectives and key terms. The objectives tell you what you'll need to know after you finish reading the section.

Key terms are listed for each section. Learn the definitions of these terms because you will most likely be tested on them. Each key term is highlighted in the text and is defined at point of use and in the margin. You can also use the glossary to locate definitions quickly.

STUDY TIP Reread the objectives and the definitions to the key terms when studying for a test to be sure you know the material.

Get Organized

A Reading Strategy at the beginning of every section provides tips to help you organize and remember the information covered in the section. Keep a science notebook so that you are ready to take notes when your teacher reviews the material in class. Keep your assignments in this notebook so that you can review them when studying for the chapter test.

SECTION 1

Earth's Oceans

What makes Earth so different from Mars? What does Earth have that Mercury doesn't?

Earth stands out from the other planets in our solar system primarily for one reason—71% of the Earth's surface is covered with water. Most of Earth's water is found in the global ocean. The global ocean is divided by the continents into four main oceans. The divisions of the global ocean are shown in **Figure 1.** The ocean is a unique body of water that plays many parts in regulating Earth's environment.

Divisions of the Global Ocean

The largest ocean is the *Pacific Ocean.* It flows between Asia and the Americas. The volume of the *Atlantic Ocean,* the second-largest ocean, is about half the volume of the Pacific. The *Indian Ocean* is the third-largest ocean. The *Arctic Ocean* is the smallest ocean. This ocean is unique because much of its surface is covered by ice. Therefore, the Arctic Ocean has not been fully explored.

Figure 1 *The global ocean is divided by the continents into four main oceans.*

READING WARM-UP

Objectives
- List the major divisions of the global ocean.
- Describe the history of Earth's oceans.
- Identify the properties of ocean water.
- Describe the interactions between the ocean and the atmosphere.

Terms to Learn

salinity
water cycle

READING STRATEGY

Discussion Read this section silently. Write down questions that you have about this section. Discuss your questions in a small group.

Arctic Ocean

Atlantic Ocean

Pacific Ocean

Indian Ocean

374   Chapter 13

Be Resourceful—Use the Web

Internet Connect
boxes in your textbook take you to resources that you can use for science projects, reports, and research papers. Go to scilinks.org, and type in the SciLinks code to get information on a topic.

Visit go.hrw.com
Find worksheets, **Current Science** magazine articles online, and other materials that go with your textbook at **go.hrw.com.** Click on the textbook icon and the table of contents to see all of the resources for each chapter.

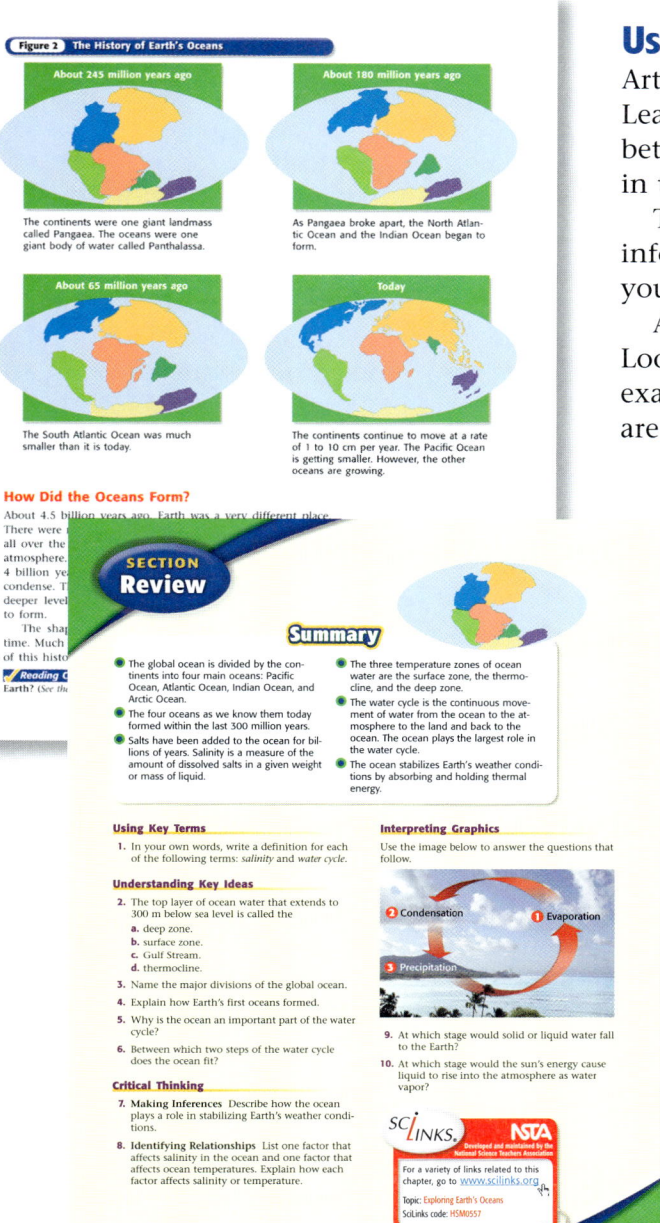

Use the Illustrations and Photos

Art shows complex ideas and processes. Learn to analyze the art so that you better understand the material you read in the text.

Tables and graphs display important information in an organized way to help you see relationships.

A picture is worth a thousand words. Look at the photographs to see relevant examples of science concepts that you are reading about.

Answer the Section Reviews

Section Reviews test your knowledge of the main points of the section. Critical Thinking items challenge you to think about the material in greater depth and to find connections that you infer from the text.

STUDY TIP When you can't answer a question, reread the section. The answer is usually there.

Do Your Homework

Your teacher may assign worksheets to help you understand and remember the material in the chapter.

STUDY TIP Don't try to answer the questions without reading the text and reviewing your class notes. A little preparation up front will make your homework assignments a lot easier. Answering the items in the Chapter Review will help prepare you for the chapter test.

Visit Holt Online Learning

If your teacher gives you a special password to log onto the Holt Online Learning site, you'll find your complete textbook on the Web. In addition, you'll find some great learning tools and practice quizzes. You'll be able to see how well you know the material from your textbook.

Visit CNN Student News

You'll find up-to-date events in science at **cnnstudentnews.com.**

SAFETY FIRST!

Exploring, inventing, and investigating are essential to the study of science. However, these activities can also be dangerous. To make sure that your experiments and explorations are safe, you must be aware of a variety of safety guidelines. You have probably heard of the saying, "It is better to be safe than sorry." This is particularly true in a science classroom where experiments and explorations are being performed. Being uninformed and careless can result in serious injuries. Don't take chances with your own safety or with anyone else's.

The following pages describe important guidelines for staying safe in the science classroom. Your teacher may also have safety guidelines and tips that are specific to your classroom and laboratory. Take the time to be safe.

Safety Rules!

Start Out Right

Always get your teacher's permission before attempting any laboratory exploration. Read the procedures carefully, and pay particular attention to safety information and caution statements. If you are unsure about what a safety symbol means, look it up or ask your teacher. You cannot be too careful when it comes to safety. If an accident does occur, inform your teacher immediately regardless of how minor you think the accident is.

Safety Symbols

All of the experiments and investigations in this book and their related worksheets include important safety symbols to alert you to particular safety concerns. Become familiar with these symbols so that when you see them, you will know what they mean and what to do. It is important that you read this entire safety section to learn about specific dangers in the laboratory.

If you are instructed to note the odor of a substance, wave the fumes toward your nose with your hand. Never put your nose close to the source.

Eye protection

Clothing protection

Hand safety

Heating safety

Electric safety

Chemical safety

Animal safety

Sharp object

Plant safety

Eye Safety

Wear safety goggles when working around chemicals, acids, bases, or any type of flame or heating device. Wear safety goggles any time there is even the slightest chance that harm could come to your eyes. If any substance gets into your eyes, notify your teacher immediately and flush your eyes with running water for at least 15 minutes. Treat any unknown chemical as if it were a dangerous chemical. Never look directly into the sun. Doing so could cause permanent blindness.

Avoid wearing contact lenses in a laboratory situation. Even if you are wearing safety goggles, chemicals can get between the contact lenses and your eyes. If your doctor requires that you wear contact lenses instead of glasses, wear eye-cup safety goggles in the lab.

Safety Equipment

Know the locations of the nearest fire alarms and any other safety equipment, such as fire blankets and eyewash fountains, as identified by your teacher, and know the procedures for using the equipment.

Neatness

Keep your work area free of all unnecessary books and papers. Tie back long hair, and secure loose sleeves or other loose articles of clothing, such as ties and bows. Remove dangling jewelry. Don't wear open-toed shoes or sandals in the laboratory. Never eat, drink, or apply cosmetics in a laboratory setting. Food, drink, and cosmetics can easily become contaminated with dangerous materials.

Certain hair products (such as aerosol hair spray) are flammable and should not be worn while working near an open flame. Avoid wearing hair spray or hair gel on lab days.

Sharp/Pointed Objects

Use knives and other sharp instruments with extreme care. Never cut objects while holding them in your hands. Place objects on a suitable work surface for cutting.

Be extra careful when using any glassware. When adding a heavy object to a graduated cylinder, tilt the cylinder so that the object slides slowly to the bottom.

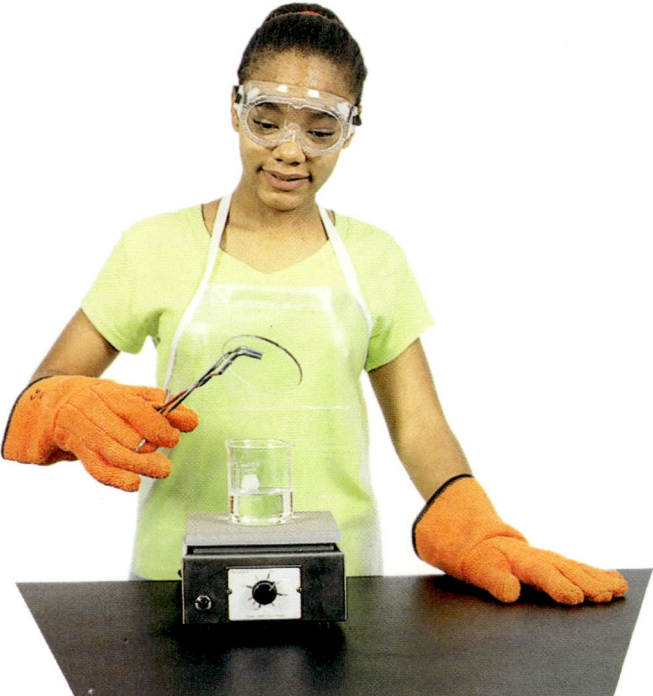

Heat

Wear safety goggles when using a heating device or a flame. Whenever possible, use an electric hot plate as a heat source instead of using an open flame. When heating materials in a test tube, always angle the test tube away from yourself and others. To avoid burns, wear heat-resistant gloves whenever instructed to do so.

Electricity

Be careful with electrical cords. When using a microscope with a lamp, do not place the cord where it could trip someone. Do not let cords hang over a table edge in a way that could cause equipment to fall if the cord is accidentally pulled. Do not use equipment with damaged cords. Be sure that your hands are dry and that the electrical equipment is in the "off" position before plugging it in. Turn off and unplug electrical equipment when you are finished.

Chemicals

Wear safety goggles when handling any potentially dangerous chemicals, acids, or bases. If a chemical is unknown, handle it as you would a dangerous chemical. Wear an apron and protective gloves when you work with acids or bases or whenever you are told to do so. If a spill gets on your skin or clothing, rinse it off immediately with water for at least 5 minutes while calling to your teacher.

Never mix chemicals unless your teacher tells you to do so. Never taste, touch, or smell chemicals unless you are specifically directed to do so. Before working with a flammable liquid or gas, check for the presence of any source of flame, spark, or heat.

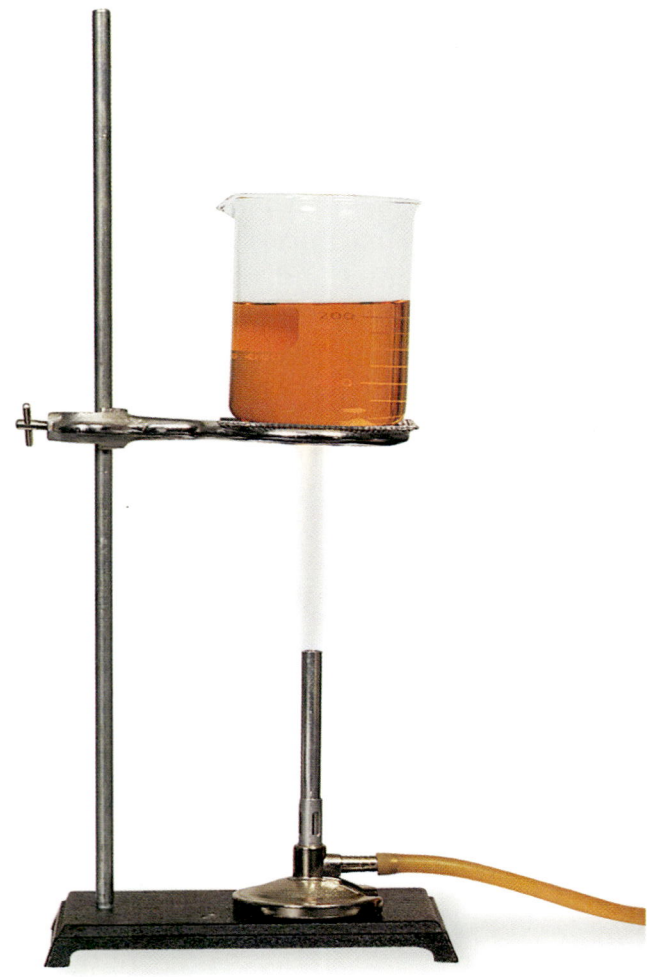

Animal Safety

Always obtain your teacher's permission before bringing any animal into the school building. Handle animals only as your teacher directs. Always treat animals carefully and respectfully. Wash your hands thoroughly after handling any animal.

Plant Safety

Do not eat any part of a plant or plant seed used in the laboratory. Wash your hands thoroughly after handling any part of a plant. When in nature, do not pick any wild plants unless your teacher instructs you to do so.

Glassware

Examine all glassware before use. Be sure that glassware is clean and free of chips and cracks. Report damaged glassware to your teacher. Glass containers used for heating should be made of heat-resistant glass.

Compression guide:
To shorten instruction because of time limitations, omit Section 2.

OBJECTIVES	LABS, DEMONSTRATIONS, AND ACTIVITIES	TECHNOLOGY RESOURCES
PACING • 90 min pp. 2–11 **Chapter Opener**	**SE Start-up Activity,** p. 3 ◆ GENERAL	**OSP Parent Letter** ■ GENERAL **CD Student Edition on CD-ROM** **CD Guided Reading Audio CD** ■ **TR Chapter Starter Transparency*** **VID Brain Food Video Quiz**
Section 1 The Active River • Describe how moving water shapes the surface of the Earth by the process of erosion. • Explain how water moves through the water cycle. • Describe a watershed. • Explain three factors that affect the rate of stream erosion. • Identify four ways that rivers are described.	**TE Demonstration** Streams Carry Fertile Sediment, p. 4 ◆ GENERAL **TE Activity** Water Re-Cycle, p. 5 BASIC **SE School-to-Home Activity** Floating Down the River, p. 6 GENERAL **TE Group Activity** Mapping River Systems, p. 6 GENERAL **TE Connection Activity** Real World, p. 7 GENERAL **TE Group Activity** Stream Load, p. 8 ◆ BASIC **TE Activity** River Field Guide, p. 8 ADVANCED **SE Connection to Language Arts** Huckleberry Finn, p. 9 GENERAL **TE Activity** Illustrating River Stages, p. 9 GENERAL **TE Connection Activity** History, p. 9 ADVANCED **SE Model-Making Labs** Water Cycle—What Goes Up…, p. 28 GENERAL	**CRF Lesson Plans*** **TR Bellringer Transparency*** **TR** The Water Cycle* **TR LINK TO LIFE SCIENCE** Rivers* **VID Lab Videos for Earth Science** **CD Science Tutor**
PACING • 45 min pp. 12–15 **Section 2 Stream and River Deposits** • Describe the four different types of stream deposits. • Describe how the deposition of sediment affects the land.	**TE Demonstration** Modeling Deposition, p. 12 ◆ GENERAL **TE Group Activity** Gold Rush, p. 13 ADVANCED **SE Science in Action** Math, Social Studies, and Language Arts Activities, pp. 34–35 GENERAL **LB Long-Term Projects & Research Ideas** Canyon Controversy* ADVANCED	**CRF Lesson Plans*** **TR Bellringer Transparency*** **CD Science Tutor**
PACING • 45 min pp. 16–21 **Section 3 Water Underground** • Identify and describe the location of the water table. • Describe an aquifer. • Explain the difference between a spring and a well. • Explain how caves and sinkholes form as a result of erosion and deposition.	**TE Demonstration** Groundwater Model, p. 16 ◆ GENERAL **TE Connection Activity** Real World, p. 17 ADVANCED **TE Group Activity** Labeling Storm Drains, p. 17 ◆ GENERAL **SE School-to-Home Activity** Water Conservation, p. 18 GENERAL **SE Connection to Environmental Science** Bat Environmentalists, p. 20 GENERAL	**CRF Lesson Plans*** **TR Bellringer Transparency*** **TR** The Water Table and Wells* **CRF SciLinks Activity*** GENERAL **SE Internet Activity** p. 17 GENERAL **CD Science Tutor**
PACING • 45 min pp. 22–27 **Section 4 Using Water Wisely** • Identify two forms of water pollution. • Explain how the properties of water influence the health of a water system. • Describe two ways that wastewater can be treated. • Describe how water is used and how water can be conserved in industry, in agriculture, and at home.	**SE Quick Lab** Measuring Alkalinity, p. 23 GENERAL **TE Group Activity** Putting Pollution in its Place, p. 24 ADVANCED **TE Connection Activity** Math, p. 25 GENERAL **TE Group Activity** A Call for Conservation, p. 26 GENERAL **SE Skills Practice Lab** Clean Up Your Act, p. 108 GENERAL **LB EcoLabs & Field Activities** The Frogs Are Off Course* ◆ ADVANCED	**CRF Lesson Plans*** **TR Bellringer Transparency*** **CD Interactive Explorations CD-ROM** Flood Bank GENERAL **CD Science Tutor**

PACING • 90 min

CHAPTER REVIEW, ASSESSMENT, AND STANDARDIZED TEST PREPARATION

CRF Vocabulary Activity* GENERAL
SE Chapter Review, pp. 30–31 GENERAL
CRF Chapter Review* ■ GENERAL
CRF Chapter Tests A* ■ GENERAL, **B*** ADVANCED, **C*** SPECIAL NEEDS
SE Standardized Test Preparation, pp. 32–33 GENERAL
CRF Standardized Test Preparation* GENERAL
CRF Performance-Based Assessment* GENERAL
OSP Test Generator GENERAL
CRF Test Item Listing* GENERAL

Online and Technology Resources

Visit **go.hrw.com** for a variety of free resources related to this textbook. Enter the keyword **HZ5DEP.**

Holt Online Learning

Students can access interactive problem-solving help and active visual concept development with the *Holt Science and Technology* Online Edition available at **www.hrw.com.**

 Guided Reading Audio CD
Also in Spanish

A direct reading of each chapter for auditory learners, reluctant readers, and Spanish-speaking students.

 Science Tutor CD-ROM

Excellent for remediation and test practice.

SKILLS DEVELOPMENT RESOURCES	SECTION REVIEW AND ASSESSMENT	STANDARDS CORRELATIONS
SE Pre-Reading Activity, p. 2 GENERAL **OSP** Science Puzzlers, Twisters & Teasers GENERAL		National Science Education Standards SAI 1
CRF Directed Reading A* ■ BASIC, B* SPECIAL NEEDS **CRF** Vocabulary and Section Summary* ■ GENERAL **SE** Reading Strategy Reading Organizer, p. 4 GENERAL **TE** Math Practice Calculating a Stream's Gradient, p. 7 GENERAL **TE** Reading Strategy Prediction Guide, p. 9 GENERAL **MS** Math Skills for Science Checking Division with Multiplication* GENERAL **SE** Connection to Language Arts Huckleberry Finn, p. 9 GENERAL	**SE** Reading Checks, pp. 4, 6, 7, 9, 10 GENERAL **TE** Homework, p. 5 ADVANCED **TE** Reteaching, p. 10 BASIC **TE** Quiz, p. 10 GENERAL **TE** Alternative Assessment, p. 10 GENERAL **SE** Section Review,* p. 11 GENERAL **CRF** Section Quiz* ■ GENERAL	UCP 1, 2, 5; ES 1c, 1f; *Chapter Lab:* SAI 1
CRF Directed Reading A* ■ BASIC, B* SPECIAL NEEDS **CRF** Vocabulary and Section Summary* ■ GENERAL **SE** Reading Strategy Prediction Guide, p. 12 GENERAL **CRF** Reinforcement Worksheet Fresh Water in the United States* BASIC **CRF** SciLinks Activity GENERAL	**SE** Reading Checks, pp. 13, 15 GENERAL **TE** Reteaching, p. 14 BASIC **TE** Quiz, p. 14 GENERAL **TE** Alternative Assessment, p. 14 GENERAL **SE** Section Review,* p. 15 GENERAL **CRF** Section Quiz* ■ GENERAL	SPSP 3, 4; ES 1c
CRF Directed Reading A* ■ BASIC, B* SPECIAL NEEDS **CRF** Vocabulary and Section Summary* ■ GENERAL **SE** Reading Strategy Discussion, p. 16 GENERAL **TE** Inclusion Strategies, p. 18 **CRF** Reinforcement Worksheet Dig It!* GENERAL **CRF** Critical Thinking Water Crisis at Happy Acres* ADVANCED	**SE** Reading Checks, pp. 16, 18, 19, 20 GENERAL **TE** Reteaching, p. 20 BASIC **TE** Quiz, p. 20 GENERAL **TE** Alternative Assessment, p. 20 GENERAL **SE** Section Review,* p. 21 GENERAL **CRF** Section Quiz* ■ GENERAL	ST 1; SPSP 3; ES 1c
CRF Directed Reading A* ■ BASIC, B* SPECIAL NEEDS **CRF** Vocabulary and Section Summary* ■ GENERAL **SE** Reading Strategy Paired Summarizing, p. 22 GENERAL **SE** Math Practice Agriculture in Israel, p. 26 GENERAL **MS** Math Skills for Science Multiplying Whole Numbers* GENERAL	**SE** Reading Checks, pp. 22, 25, 26, 27 GENERAL **TE** Homework, p. 23 GENERAL **TE** Reteaching, p. 26 BASIC **TE** Quiz, p. 26 GENERAL **TE** Alternative Assessment, p. 26 GENERAL **SE** Section Review,* p. 27 GENERAL **CRF** Section Quiz* ■ GENERAL	UCP 2, 3; SAI 1; SPSP 3; *LabBook:* UCP 2, 3; ST 2; SAI 1; SPSP 5

One-Stop Planner® CD-ROM

This convenient CD-ROM includes:
- Lab Materials QuickList Software
- Holt Calendar Planner
- Customizable Lesson Plans
- Printable Worksheets
- ExamView® Test Generator

cnnstudentnews.com

Find the latest news, lesson plans, and activities related to important scientific events.

www.scilinks.org

Maintained by the **National Science Teachers Association.** See Chapter Enrichment pages for a complete list of topics.

Check out *Current Science* articles and activities by visiting the HRW Web site at **go.hrw.com.** Just type in the keyword **HZ5CS11T.**

 **Classroom Videos**

- **Lab Videos** demonstrate the chapter lab.
- **Brain Food Video Quizzes** help students review the chapter material.
- **CNN Videos** bring science into your students' daily life.

Visual Resources

CHAPTER STARTER TRANSPARENCY

BELLRINGER TRANSPARENCIES

Section: The Active River
The source of the Mississippi River is said to be Lake Itasca in Northern Minnesota. Is Lake Itasca really the ultimate source of the Mississippi River? Can a river really be said to have a source? Describe what a river's source might look like.

Write your response in your **science journal**.

Section: Stream and River Deposits
The Nile River Delta floods frequently, but many farms are located on the delta. In fact, even though flooding along rivers is potentially dangerous, many farmers have traditionally located their farms near rivers. Is this a sound agricultural practice? Where would you locate a farm?

Write your answer in your **science journal**.

TEACHING TRANSPARENCIES

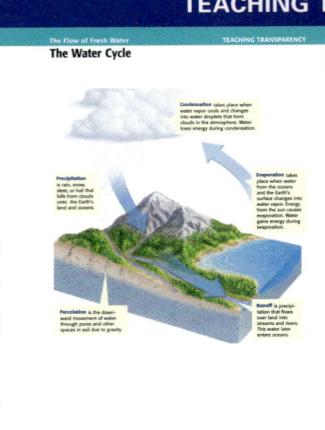

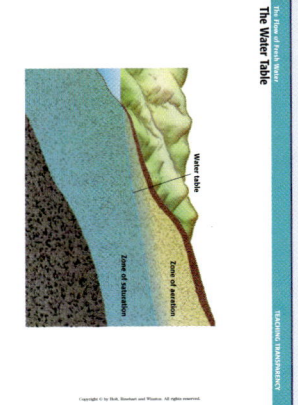

TEACHING TRANSPARENCIES

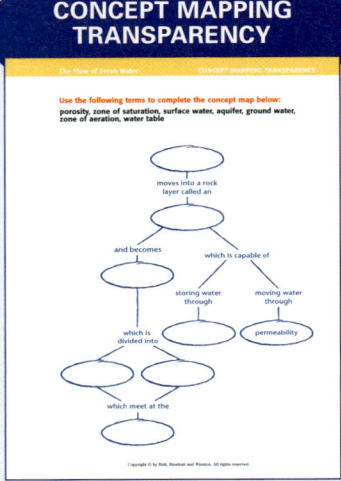

CONCEPT MAPPING TRANSPARENCY

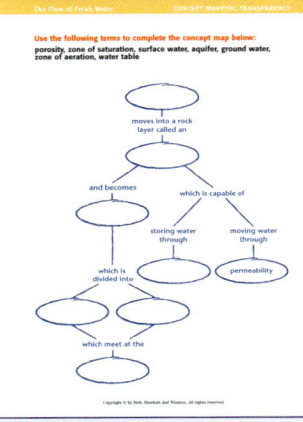

Planning Resources

LESSON PLANS

Lesson Plan — SAMPLE

Section: Waves

Pacing
Regular Schedule: with lab(s):2 days · without lab(s):2 days
Block Schedule: with lab(s):1 1/2 days · without lab(s):1 day

Objectives
1. Relate the seven properties of life to a living organism.
2. Describe seven themes that can help you to organize what you learn about biology.
3. Identify the tiny structures that make up all living organisms.
4. Differentiate between reproduction and heredity and between metabolism and homeostasis.

National Science Education Standards Covered
LSInter6:Cells have particular structures that underlie their functions.
LSMat1:Most cell functions involve chemical reactions.
LSBeh1:Cells store and use information to guide their functions.
UCP1:Cell functions are regulated.
SI1: Cells can differentiate and form complete multicellular organisms.
PS1: Species evolve over time.
ESS1: The great diversity of organisms is the result of more than 3.5 billion years of evolution.
ESS2: Natural selection and its evolutionary consequences provide a scientific explanation for the fossil record of ancient life forms as well as for the striking molecular similarities observed among the diverse species of living organisms.
ST1: The millions of different species of plants, animals, and microorganisms that live on Earth today are related by descent from common ancestors.
ST2: The energy for life primarily comes from the sun.
SPSP1: The complexity and organization of organisms accommodates the need for obtaining, transforming, transporting, releasing, and eliminating the matter and energy used to sustain the organism.
SPSP6: As matter and energy flows through different levels of organization of living systems—cells, organs, communities—and between living systems and the physical environment, chemical elements are recombined in different ways.
HNS1: Organisms have behavioral responses to internal changes and to external stimuli.

PARENT LETTER

SAMPLE

Dear Parent,

Your son's or daughter's science class will soon begin exploring the chapter entitled "The World of Physical Science." In this chapter, students will learn about how the scientific method applies to the world of physical science and the role of physical science in society. By the end of the chapter, students should demonstrate a clear understanding of the chapter's main ideas and be able to discuss the following topics:

1. physical science is the study of energy and matter (Section 1)
2. the role of physical science in the world around them (Section 1)
3. careers that rely on physical science (Section 1)
4. the steps used in the scientific method (Section 2)
5. examples of technology (Section 2)
6. how the scientific method is used to answer questions and solve problems (Section 2)
7. how our knowledge of science changes over time (Section 2)
8. how models represent real objects or systems (Section 3)
9. examples of different ways models are used in science (Section 3)
10. the importance of the International System of Units (Section 4)
11. the appropriate units to use for particular measurements (Section 4)
12. how area and density are derived quantities (Section 4)

Questions to Ask Along the Way

You can help your son or daughter learn about these topics by asking interesting questions such as the following:

• What are some surprising careers that use physical science?
• What is a characteristic of a good hypothesis?
• When is it a good idea to use a model?
• Why do Americans measure things in terms of inches and yards and meters ?

ALSO IN SPANISH

TEST ITEM LISTING

TEST ITEM LISTING
The World of Earth Science — SAMPLE

MULTIPLE CHOICE

1. A limitation of models is that
   a. they are large enough to see.
   b. they do not act exactly like the things that they model.
   c. they are smaller than the things that they model.
   d. they model unfamiliar things.
   Answer: B   Difficulty: 1   Section: 3   Objective: 2
2. The length 10 m is equal to
   a. 100 cm.   c. 10,000 mm.
   b. 1,000 cm.   d. Both (b) and (c)
   Answer: B   Difficulty: 1   Section: 3   Objective: 2
3. To be valid, a hypothesis must be
   a. testable.   c. made into a law.
   b. supported by evidence.   d. Both (a) and (b)
   Answer: D   Difficulty: 1   Section: 2   Objective: 2 1
4. The statement "Sheila has a stain on her shirt" is an example of a(n)
   a. law.   c. observation.
   b. hypothesis.   d. prediction.
   Answer: B   Difficulty: 1   Section: 2   Objective: 2
5. A hypothesis is often developed out of
   a. observations.   c. laws.
   b. experiments.   d. Both (a) and (b)
   Answer: B   Difficulty: 1   Section: 2   Objective: 2
6. How many milliliters are in 3.5 kL?
   a. 3500. mL   c. 7,500, 000 mL
   b. 0.0035 mL   d. 35,000 mL.
   Answer: B   Difficulty: 1   Section: 3
7. A map of Seattle is an example of a
   a. law.   c. model.
   b. theory.   d. unit.
   Answer: B   Difficulty: 1   Section: 2   Objective: 2
8. A lab has the safety icons shown below. These icons mean that you should wear
   a. safety goggles.   c. safety goggles and a lab apron.
   b. only a lab apron.   d. safety goggles, a lab apron, and gloves.
   Answer: B   Difficulty: 1   Section: 3   Objective: 2
9. The law of conservation of mass says the tot al mass before a chemical change is
   a. more than the total mass after the change.
   b. less than the total mass after the change.
   c. the same as the total mass after the change.
   d. not the same as the total mass after the change.
   Answer: B   Difficulty: 1   Section: 3
10. In which of the following ones might you find a geochemist at work?
   a. studying the chemistry of rocks   c. studying fishes
   b. studying forestry   d. studying the atmosphere
   Answer: B   Difficulty: 1   Section: 3   Objective: 2

One-Stop Planner® CD-ROM

This CD-ROM includes all of the resources shown here and the following time-saving tools:

• *Lab Materials QuickList Software*
• *Customizable lesson plans*
• *Holt Calendar Planner*
• *The powerful ExamView® Test Generator*

Meeting Individual Needs

DIRECTED READING A
Directed Reading A — SAMPLE
Section: THAT'S SCIENCE!
1. How did James Czarnowski get his idea for the penguin boat, Proteus? Explain.
BASIC
ALSO IN SPANISH

DIRECTED READING B
Directed Reading B — SAMPLE
Section: THAT'S SCIENCE!
1. How did James Czarnowski get his idea for the penguin boat, Proteus? Explain.
2. What is unusual about the way that Proteus moves through the water?
SPECIAL NEEDS

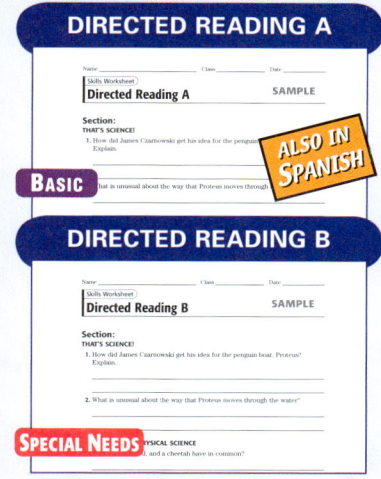

VOCABULARY ACTIVITY
Vocabulary Activity — SAMPLE
Getting the Dirt on the Soil
After you finish reading Chapter [Unique Title], try this puzzle! Use the clues below to unscramble the vocabulary words.
GENERAL

VOCABULARY AND SECTION SUMMARY
Vocabulary & Notes — SAMPLE
Section: VOCABULARY
In your own words, write a definition of the following term in the space provided.
1. scientific method
2. technology
GENERAL
ALSO IN SPANISH

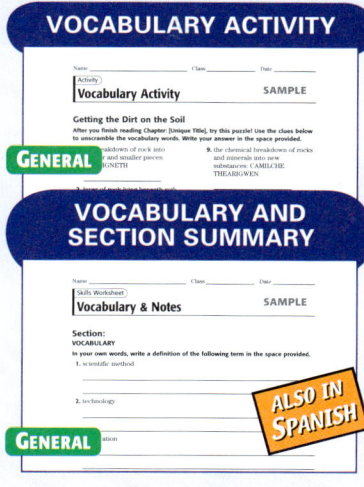

REINFORCEMENT
Reinforcement — SAMPLE
The Plane Truth
Complete this worksheet after you finish reading the Section: [Unique Section Title]
BASIC

CRITICAL THINKING
Critical Thinking — SAMPLE
A Solar Solution
ADVANCED

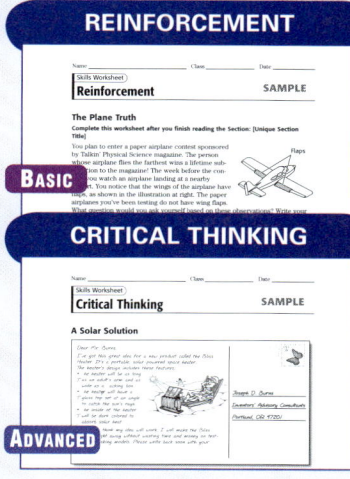

SCILINKS ACTIVITY
SciLinks Activity — SAMPLE
MARINE ECOSYSTEMS
Go to www.scilinks.com. To find links related to marine ecosystems, type in the keyword HL5###. Then, use the links to answer the questions about marine ecosystems.
GENERAL

SCIENCE PUZZLERS, TWISTERS & TEASERS
SCIENCE PUZZLERS, TWISTERS & TEASERS — CHAPTER 11
The Flow of Fresh Water
Map to a Message
Follow the instructions on this worksheet to uncover a hidden message.
1. Use the following clues to fill in the blanks with terms from the chapter.
   a. Where aeration and saturation zones meet
   b. When rain falls into a stream
   c. A stream follows this
   d. output forms this deposit
   e. If a rock is this, water can flow through it
GENERAL

Labs and Activities

ECOLABS & FIELD ACTIVITIES
STUDENT WORKSHEET 12 — DISCOVERY LAB
The Frogs Are Off Course
Extra! Extra! Trumpet County Frog Numbers Are Dwindling!
ADVANCED

LONG-TERM PROJECTS & RESEARCH IDEAS
PROJECT 39 — STUDENT WORKSHEET — DESIGN YOUR OWN
Canyon Controversy
ADVANCED

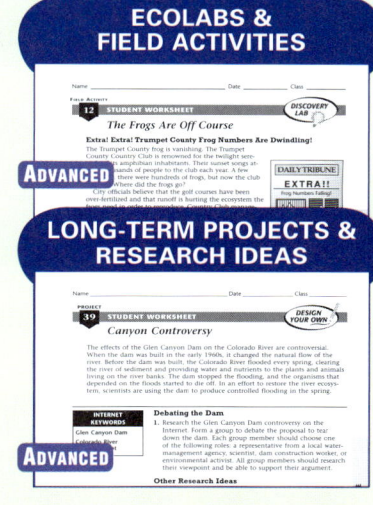

DATASHEETS FOR QUICKLABS
TEACHER RESOURCE PAGE
Quick Lab — DATASHEET FOR QUICK LAB
Reaction to Stress — SAMPLE
Background

DATASHEETS FOR CHAPTER LABS
TEACHER RESOURCE PAGE
Skills Practice Lab — DATASHEET FOR CHAPTER LAB
Using Scientific Methods — SAMPLE
Teacher's Notes
TIME REQUIRED
One 45-minute class period.

DATASHEETS FOR LABBOOK
TEACHER RESOURCE PAGE
Skills Practice Lab — DATASHEET FOR LABBOOK LAB
Does It All Add Up? — SAMPLE
Teacher's Notes
TIME REQUIRED
One 45-minute class period.

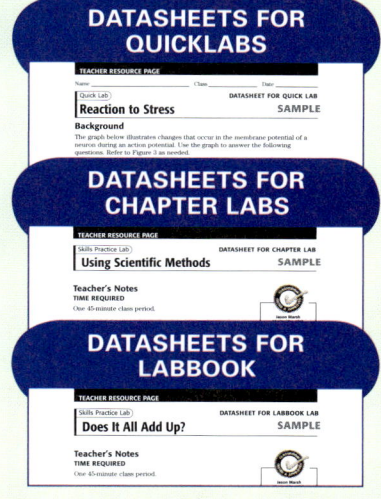

Review and Assessments

SECTION QUIZ
Section Quiz — SAMPLE
Section:
In the space provided, write the letter of the description that best matches the term or phrase.
_____ 1. building molecules that can be used as an energy source, or breaking down molecules in which energy is stored
GENERAL
ALSO IN SPANISH

SECTION REVIEW
Section Review — SAMPLE
Section:
KEY TERMS
1. What do paleontologists study?
2. How does a trace fossil differ from petrified wood?
GENERAL
ALSO IN SPANISH

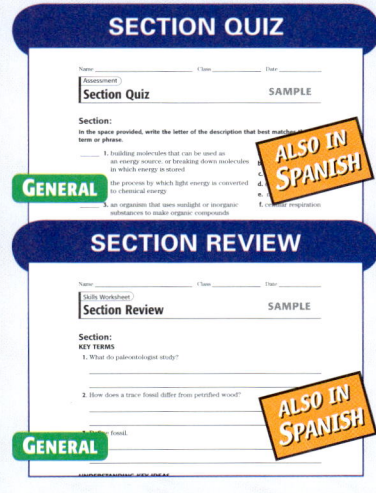

CHAPTER REVIEW
Chapter Review — SAMPLE
USING VOCABULARY
1. Define biome in your own words.
2. Describe the characteristics of a savanna and a desert.
GENERAL

CHAPTER TEST A
Chapter Test A — SAMPLE
MULTIPLE CHOICE
In the space provided, write the letter of the term or phrase that best completes each statement or best answers each question.
_____ 1. Surface currents are formed by
   a. the moon's gravity.   c. wind.
   b. the sun's gravity.   d. increased water density.
_____ 2. When waves come near the shore,
   a. they speed up.   c. their wavelength increases.
   b. they maintain their speed. d. their wave height increases.
GENERAL
ALSO IN SPANISH

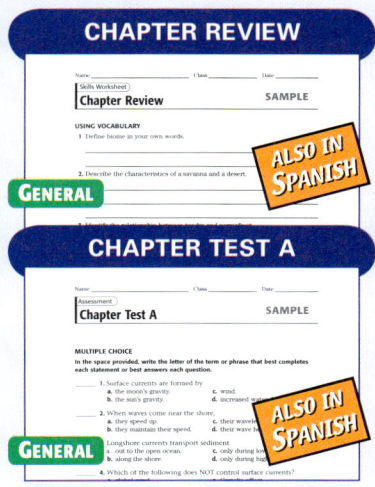

CHAPTER TEST B
Chapter Test B — SAMPLE
MULTIPLE CHOICE
In the space provided, write the letter of the term or phrase that best completes each statement or best answers each question.
_____ 1. Surface currents are formed by
   a. the moon's gravity.   c. wind.
   b. the sun's gravity.   d. increased water density.
_____ 2. When waves come near the shore,
   a. they speed up.   c. their wavelength increases.
   b. they maintain their speed. d. their wave height increases.
ADVANCED

CHAPTER TEST C
Chapter Test C — SAMPLE
MULTIPLE CHOICE
In the space provided, write the letter of the term or phrase that best completes each statement or best answers each question.
_____ 1. Surface currents are formed by
   a. the moon's gravity.   c. wind.
   b. the sun's gravity.   d. increased water density.
_____ 2. When waves come near the shore,
   a. they speed up.   c. their wavelength increases.
   b. they maintain their speed. d. their wave height increases.
SPECIAL NEEDS

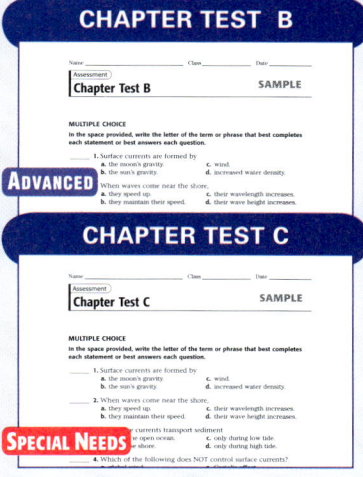

STANDARDIZED TEST PREPARATION
Standardized Test Preparation — SAMPLE
READING
Read the passages below. Then, read each question that follows the passage. Decide which is the best answer to each question.
GENERAL

PERFORMANCE-BASED ASSESSMENT
Performanced-Based Assessment — SAMPLE — SKILL BUILDER
OBJECTIVE
Determine which factors cause some sugar shapes to break down faster than others.
KNOW THE SCORE!
As you work through the activity, keep in mind that you will be earning a grade for the following
• how you form and test the hypothesis (30%)
• the quality of your analysis (40%)
• the clarity of your conclusions (30%)
Using Scientific Methods
QUESTIONS
MATERIALS AND EQUIPMENT
GENERAL

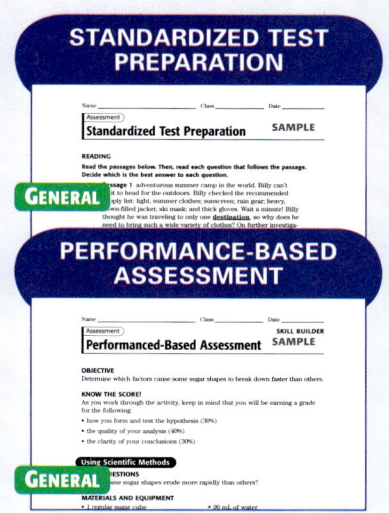

This Chapter Enrichment provides relevant and interesting information to expand and enhance your presentation of the chapter material.

Section 1

The Active River

William Morris Davis

- William Morris Davis (1850–1934) was a famous American geographer who was the first to propose the erosion cycle. Davis theorized that landscapes are initially uplifted. Streams flow rapidly from the uplifted land and cut into the landscape. Gradually, the landscape's slope is reduced. Eventually, the landscape changes into an old erosional surface that is fairly flat. Davis's theory is not supported by scientists today, however. The process of river erosion is approached today from a systems perspective. Each system is composed of different parts that vary from landscape to landscape.

Stream Flow

- Streams have two general types of flow—laminar and turbulent. Laminar flow occurs when the stream load moves in a generally parallel flow. This movement occurs where the channel is smooth. Streams with a greater velocity experience turbulent flow. The stream load generally is rolled, lifted, and bounced along, causing much more erosion in the stream channel.

Section 2

Stream and River Deposits

Drainage Patterns

- A drainage pattern is the arrangement of river channels in a drainage basin. A drainage pattern is determined by an area's geology and climate. One of the most

common patterns, called *dendritic,* is a treelike pattern that forms where rocks and sediments are flat. A *parallel* pattern forms where there are valleys and ridges. And a *radial* pattern forms when streams flow from a central peak, such as a volcanic mountain.

Deltas

- The word *delta* was first used by the Greek historian Herodotus to describe the mouth of a river. In the fifth century BCE, Herodotus was traveling in Egypt when he saw the triangular mouth of the Nile River and named the shape after the Greek letter, delta.

Is That a Fact!

- ◆ Over the last 6,000 years, the Mississippi River delta has shifted from east to west several times. Today, the river empties to the east, but scientists think that if left alone the river would change its course and head toward a swampy region called the Atchafalaya Basin. Only massive dams keep the Mississippi River on its present course. If the river channel changes, the river may no longer pass through New Orleans.

Section 3

Water Underground

Caves as Shelters

- The flow of fresh water underground created caves that were just as important to the development of human civilizations as fertile flood plains were. Humans have used caves as shelters for hundreds of thousands of years. Evidence suggests that the use of caves for shelter coincides with the first controlled use of fire. Hearths that may be 750,000 years old have been found in the cave of l'Escale, in southeastern France.

- In China, excavations in a cave called Chou-k'ou-tien have yielded 400,000-year-old fossilized remains of *Homo erectus*. Evidence of charred animal bones suggests that the inhabitants may have cooked their food.

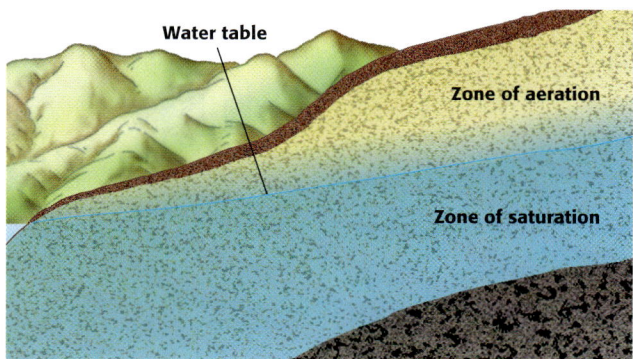

Water table
Zone of aeration
Zone of saturation

Is That a Fact!

- ◆ The largest cave chamber in the world is the Sarawak chamber, in Malaysia. The chamber is 600 m long and has an average width of 450 m.

Section 4

Using Water Wisely

Aquifers

- There are two types of aquifers. The first forms in *consolidated formations,* which are formed from solid rock overlaid with permeable rock that is saturated with water. The second kind of aquifer forms in *unconsolidated formations*—loose sand, soil, and gravel. The amount of water contained in an unconsolidated aquifer depends on how tightly the materials are packed. Because of this, sand and gravel, which are coarse-grained, are usually high-yield aquifers, while formations that are finer-grained tend to hold less water.

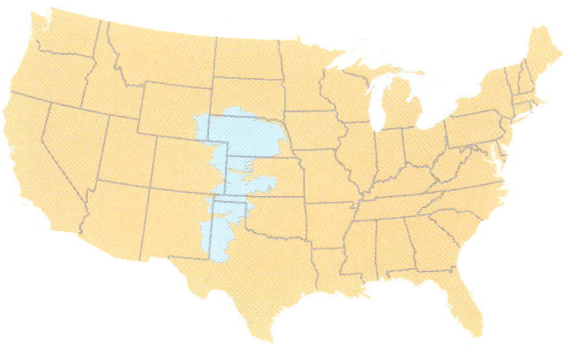

Acequias in New Mexico

- New Mexico receives very little rain each year, so water conservation is critical. Because wide irrigation ditches lose a great deal of water to evaporation, New Mexicans have relied for centuries on shallow earthen ditches fed by local rivers to supply growing plants with water. Each ditch, called an *acequia,* provides water to a small area. Acequias allow for water to seep into the ground, minimizing evaporation and allowing water to reach plant roots. Water that isn't absorbed by the soil returns to the river and provide water to people downstream to irrigate their fields.

Is That a Fact!

- ◆ When ground water is depleted so quickly that the system cannot recharge, there can be dramatic consequences. At Edwards Air Force Base, in California, the aquifer has lost water so quickly that ground settling has led to sinks and fissures. One of the fissures is about 625 m long!

SciLinks is maintained by the National Science Teachers Association to provide you and your students with interesting, up-to-date links that will enrich your classroom presentation of the chapter.

Visit www.scilinks.org and enter the SciLinks code for more information about the topic listed.

Developed and maintained by the National Science Teachers Association

Topic: Rivers and Streams
SciLinks code: HSM1316

Topic: Stream Deposits
SciLinks code: HSM1458

Topic: Water Underground
SciLinks code: HSM1633

Topic: Water Erosion
SciLinks code: HSM1627

Topic: Water Pollution and Conservation
SciLinks code: HSM1630

Overview

Tell students that this chapter will help them learn about the movement of fresh water on the Earth's surface and underground. The chapter describes the water cycle, erosion and deposition by rivers, and the characteristics of water underground. The chapter also discusses water pollution, wastewater treatment, and water conservation.

Assessing Prior Knowledge

Students should be familiar with the following topics:

• erosion and deposition
• states of matter

Identifying Misconceptions

Students may be confused about how water is stored underground. Explain to students that groundwater is usually not stored in giant underground caves. Rather, spaces between and within rocks fill with water. In this way, an aquifer is much like a sponge.

The Flow of Fresh Water

About the

You can hear the roar of Iguaçu (EE gwah SOO) Falls for miles. The Iguaçu River travels more than 500 km across Brazil before it tumbles off the edge of a volcanic plateau in a series of 275 individual waterfalls. Over the past 20,000 years, erosion has caused the falls to move 28 km upstream.

PRE-READING ACTIVITY

FOLDNOTES **Booklet** Before you read the chapter, create the FoldNote entitled "Booklet" described in the **Study Skills** section of the Appendix. Label each page of the booklet with a main idea from the chapter. As you read the chapter, write what you learn about each main idea on the appropriate page of the booklet.

Standards Correlations

National Science Education Standards

The following codes indicate the National Science Education Standards that correlate to this chapter. The full text of the standards is at the front of the book.

Chapter Opener
SAI 1

Section 1 The Active River
UCP 1, 2, 5; ES 1c, 1f

Section 2 Stream and River Deposits
ES 1c; SPSP 3, 4

Section 3 Water Underground
ST1; ES 1c; SPSP 3

Section 4 Using Water Wisely
UCP 2, 3; SAI 1; SPSP 3; *LabBook:* UCP 2, 3; ST 2; SAI 1; SPSP 5

Chapter Lab
SAI 1

Chapter Review
ES 1c, 1f

Science in Action
SPSP 5

START-UP ACTIVITY

MATERIALS

FOR EACH GROUP
• clothespin
• cup, paper
• gravel
• magnifying lens
• sand, bucket of
• washtub, plastic, rectangular
• water

Teacher's Notes: Students can tilt the tub by placing a block of wood or a book under one end.

Answers

1. Answers may vary. The moving water cut into the sand and formed a small groove.

2. Answers may vary. As time passed, the moving water cut deeper into the sand and created a wider groove.

3. Accept all reasonable responses. Sample answer: Runoff (water) moves over the land, cutting a gully into the soil. At first the gully is shallow and narrow, but over time the gully widens and becomes deeper. If there is enough water, the gully eventually becomes a river. (Students should note that the model is accurate in that moving water does produce similar landforms. The model is inaccurate for several reasons, including its scale and the fact that rivers flow over varied terrain, not just over uniform sand deposits.)

START-UP ACTIVITY

Stream Weavers

Do the following activity to learn how streams and river systems develop.

Procedure

1. Begin with enough **sand** and **gravel** to fill the bottom of a **rectangular plastic washtub.**

2. Spread the gravel in a layer at the bottom of the washtub. On top of the gravel, place a layer of sand that is 4 cm to 6 cm deep. Add more sand to one end of the washtub to form a slope.

3. Make a small hole in the bottom of a **paper cup.** Attach the cup to the inside wall of the tub with a **clothespin.** The cup should be placed at the end that has more sand.

4. Fill the cup with **water,** and observe the water as it moves over the sand. Use a **magnifying lens** to observe features of the stream more closely.

5. Record your observations.

Analysis

1. At the start of your experiment, how did the moving water affect the sand?

2. As time passed, how did the moving water affect the sand?

3. Explain how this activity modeled the development of streams. In what ways was the model accurate? How was it inaccurate?

Chapter Starter Transparency
Use this transparency to help students begin thinking about how rivers can affect the Earth's surface.

CHAPTER RESOURCES

Technology

 **Transparencies**
• Chapter Starter Transparency

 READING SKILLS

 Student Edition on CD-ROM

 Guided Reading Audio CD
• English or Spanish

Classroom Videos
• Brain Food Video Quiz

Workbooks

Science Puzzlers, Twisters & Teasers
• The Flow of Fresh Water **GENERAL**

Focus

Overview

This section introduces the water cycle and discusses the role that rivers play in the movement of fresh water. Students will learn that rivers are changing, dynamic systems that continually shape the land. The section discusses the factors that contribute to rates of stream erosion and concludes with descriptions of various types of rivers.

 Bellringer

Ask students to discuss whether a river can be said to have a source and describe what a river's source might look like.
LS Verbal/Interpersonal

Motivate

Demonstration — GENERAL

Streams Carry Fertile Sediment
Add a teaspoon of brightly colored tempera paint to half a cup of soil. Place the soil in a funnel lined with filter paper, and place the funnel over a large jar. Tell students that the paint represents nutrients in the soil. Ask students to predict what will happen when it rains. Demonstrate rain by pouring water into the funnel. Discuss the role that rivers play in distributing soil nutrients. **LS Visual**

READING WARM-UP

Objectives

- Describe how moving water shapes the surface of the Earth by the process of erosion.
- Explain how water moves through the water cycle.
- Describe a watershed.
- Explain three factors that affect the rate of stream erosion.
- Identify four ways that rivers are described.

Terms to Learn

erosion	divide
water cycle	channel
tributary	load
watershed	

READING STRATEGY

Reading Organizer As you read this section, create an outline of the section. Use the headings from the section in your outline.

---

erosion the process by which wind, water, ice, or gravity transports soil and sediment from one location to another

Figure 1 The Grand Canyon is located in northwestern Arizona. The canyon formed over millions of years as running water eroded the rock layers. (In some places, the canyon is now 29 km wide.)

The Active River

If you had fallen asleep with your toes dangling in the Colorado River 6 million years ago and you had woken up today, your toes would be hanging about 1.6 km (about 1 mi) above the river!

The Colorado River carved the Grand Canyon, shown in **Figure 1,** by washing billions of tons of soil and rock from its riverbed. The Colorado River made the Grand Canyon by a process that can take millions of years.

Rivers: Agents of Erosion

Six million years ago, the area now known as the Grand Canyon was nearly as flat as a pancake. The Colorado River cut down into the rock and formed the Grand Canyon over millions of years through a process called erosion. **Erosion** is the process by which soil and sediment are transported from one location to another. Rivers are not the only agents of erosion. Wind, rain, ice, and snow can also cause erosion.

Because of erosion caused by water, the Grand Canyon is now about 1.6 km deep and 446 km long. In this section, you will learn about stream development, river systems, and the factors that affect the rate of stream erosion.

✓ **Reading Check** Describe the process that created the Grand Canyon. (*See the Appendix for answers to Reading Checks.*)

Answer to Reading Check

The Colorado River eroded the rock over millions of years.

The Water Cycle

Have you ever wondered how rivers keep flowing? Where do rivers get their water? Learning about the water cycle, shown in **Figure 2,** will help you answer these questions. The **water cycle** is the continuous movement of Earth's water from the ocean to the atmosphere to the land and back to the ocean. The water cycle is driven by energy from the sun.

water cycle the continuous movement of water from the ocean to the atmosphere to the land and back to the ocean

Figure 2 The Water Cycle

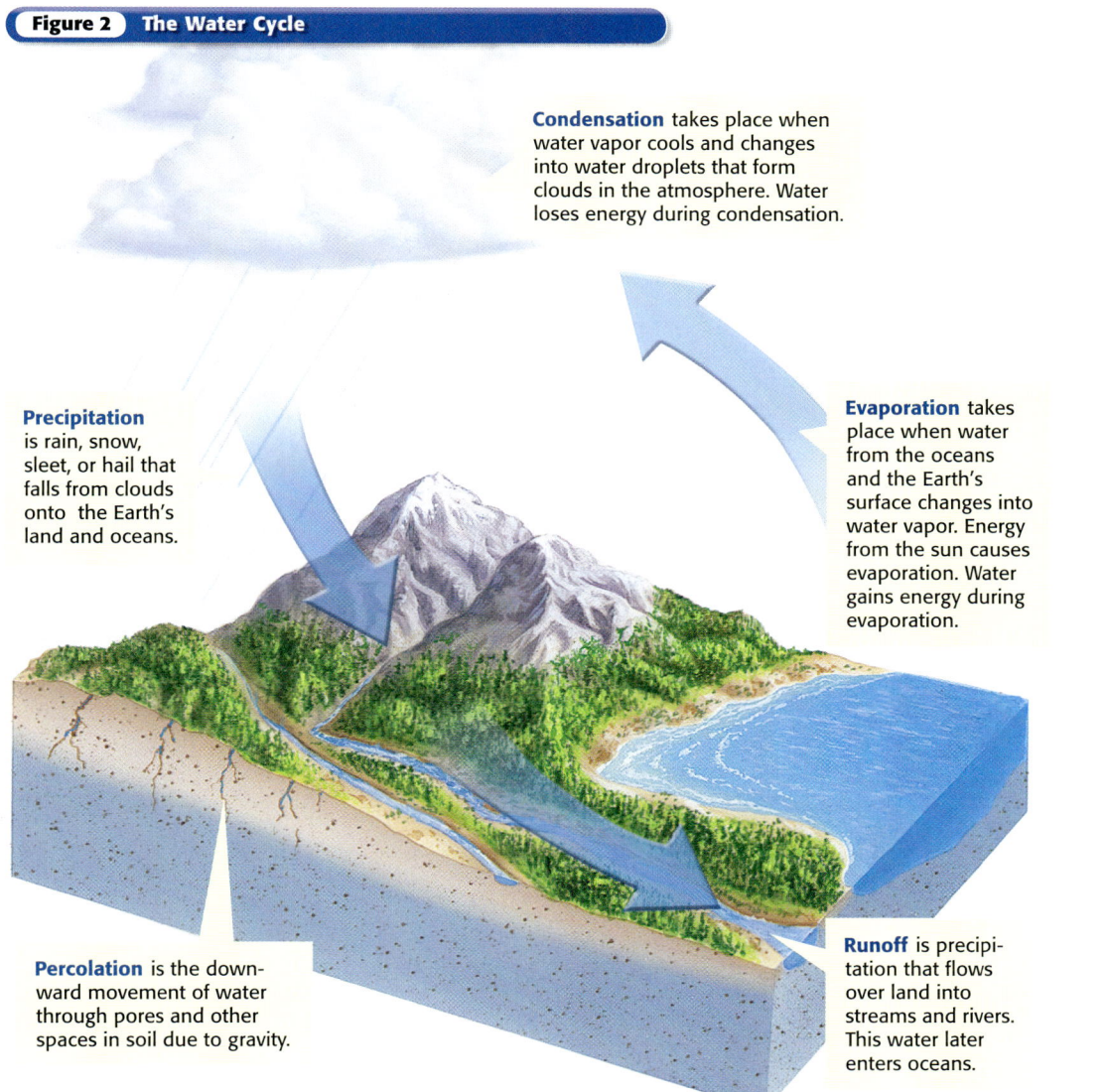

Condensation takes place when water vapor cools and changes into water droplets that form clouds in the atmosphere. Water loses energy during condensation.

Precipitation is rain, snow, sleet, or hail that falls from clouds onto the Earth's land and oceans.

Evaporation takes place when water from the oceans and the Earth's surface changes into water vapor. Energy from the sun causes evaporation. Water gains energy during evaporation.

Percolation is the downward movement of water through pores and other spaces in soil due to gravity.

Runoff is precipitation that flows over land into streams and rivers. This water later enters oceans.

Teach

ACTIVITY ——— BASIC

Water Re-Cycle Have students reproduce the water-cycle diagram and rewrite the labels in their own words. Have students exchange diagrams and review each other's work. **LS Visual**

CONNECTION to Environmental Science ——— ADVANCED

Large Dam Projects Damming and controlling rivers creates usable farmland, allows settlement in flood plains, and powers industry. Beginning in the 1930s, massive hydroelectric dams were built across many western rivers. Many developing nations are now following this example.

While dams provide many benefits, large dam projects face increasing criticism. Around the world, as many as 60 million people have been displaced by dams. Even more people have suffered from the effects of being downstream from dams: farmland is deprived of flood sediments and water for irrigation becomes unusable, fisheries become less productive, and epidemics of waterborne disease often follow large dam projects. Have students research large dam projects and debate this question: "After the industrialized world has benefited so much from damming its rivers, can industrialized nations ask developing countries not to exploit their water resources?" **LS Logical/Interpersonal**

Homework ——— ADVANCED

World River Scrapbook Beginning with early Mesopotamian cultures of the Tigris and Euphrates River valleys, river systems have been centers of human civilization. Have students choose an important world river and create a scrapbook in which they explore the cultural, ecological, and economic significance of the river. Students should apply concepts from each section as they create the scrapbooks. When they finish the chapter, have students present their scrapbooks to the class. **LS Visual/Kinesthetic**

MISCONCEPTION /// ALERT \\\

Evaporation Everywhere Students may think that water evaporates only from rivers and other bodies of water, not from soil. Display the transparency entitled "The Water Cycle." Explain to students that water also evaporates from soil during the water cycle and that plants contribute to the water cycle. They draw liquid water from the ground and transpire, or release water vapor through leaf pores.

Group ACTIVITY — GENERAL

Mapping River Systems

Have pairs of students use a map or atlas to create a poster that illustrates the river systems and drainage basins of North America. Encourage them to use arrows indicating the direction of flow in major rivers and to draw the divides that separate each drainage basin. Working independently, students could make a map that shows the river systems of your county or state. **LS Interpersonal**

SCHOOL to HOME

Floating down the River
Study a map of the United States at home with a parent. Find the Mississippi River. Imagine that you are planning a rafting trip down the river. On the map, trace the route of your trip from Lake Itasca, Minnesota to the mouth of the river in Louisiana. If you were floating on a raft down the Mississippi River, what major tributaries would you pass? What cities would you pass? Mark them on the map. How many kilometers would you travel on this trip?

River Systems

The next time you take a shower, notice that individual drops of water join together to become small streams. These streams join other small streams and form larger ones. Eventually, all of the water flows down the drain. Every time you shower, you create a model river system—a network of streams and rivers that drains an area of its runoff. Just as the shower forms a network of flowing water, streams and rivers form a network of flowing water on land. A stream that flows into a lake or into a larger stream is called a **tributary.**

Watersheds

River systems are divided into regions called watersheds. A **watershed,** or *drainage basin,* is the area of land that is drained by a water system. The largest watershed in the United States is the Mississippi River watershed. The Mississippi River watershed has hundreds of tributaries that extend from the Rocky Mountains, in the West, to the Appalachian Mountains, in the East.

The satellite image in **Figure 3** shows that the Mississippi River watershed covers more than one-third of the United States. Other major watersheds in the United States are the Columbia River, Rio Grande, and Colorado River watersheds. Watersheds are separated from each other by an area of higher ground called a **divide.**

✓ **Reading Check** Describe the difference between a watershed and a divide.

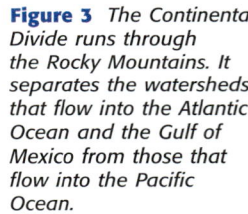

Figure 3 *The Continental Divide runs through the Rocky Mountains. It separates the watersheds that flow into the Atlantic Ocean and the Gulf of Mexico from those that flow into the Pacific Ocean.*

CONNECTION to History ——— GENERAL

Mono Reservoir In 1935, a dam was built in California to prevent sediment from filling Gibraltar Reservoir. This dam created Mono Reservoir. The watershed above Mono Reservoir was burned by forest fires, and people worried that sediment would fill the reservoirs before plants could regrow to hold the soil. Unfortunately, the next 2 years saw record rainfall. Sediment filled Mono Reservoir and half-filled Gibraltar Reservoir.

Figure 4 *A mountain stream, such as the one at left, at Kenai Peninsula in Alaska, flows rapidly and has more erosive energy. A river on a flat plain, such as the Kuskowin River in Alaska, shown below, flows slowly and has less erosive energy.*

Stream Erosion

As a stream forms, it erodes soil and rock to make a channel. A **channel** is the path that a stream follows. When a stream first forms, its channel is usually narrow and steep. Over time, the stream transports rock and soil downstream and makes the channel wider and deeper. When streams become longer and wider, they are called *rivers*. A stream's ability to erode is influenced by three factors: gradient, discharge, and load.

Gradient

Figure 4 shows two photos of rivers with very different gradients. *Gradient* is the measure of the change in elevation over a certain distance. A high gradient gives a stream or river more erosive energy to erode rock and soil. A river or stream that has a low gradient has less energy for erosion.

Discharge

The amount of water that a stream or river carries in a given amount of time is called *discharge*. The discharge of a stream increases when a major storm occurs or when warm weather rapidly melts snow. As the stream's discharge increases, its erosive energy and speed and the amount of materials that the stream can carry also increase.

✓ **Reading Check** What factors cause a stream to flow faster?

tributary a stream that flows into a lake or into a larger stream

watershed the area of land that is drained by a water system

divide the boundary between drainage areas that have streams that flow in opposite directions

channel the path that a stream follows

Calculating a Stream's Gradient

If a stream starts at an elevation of 4,900 m and travels 450 km downstream to a lake that is at an elevation of 400 m, what is the stream's gradient? (Hint: Subtract the final elevation from the starting elevation, and divide by 450. Don't forget to keep track of the units.)

Group ACTIVITY — BASIC

Stream Load To help students learn the differences between the types of loads that streams carry, have them do the activity in groups of four. Have groups fill a plastic jar three-quarters full of water. Provide a few small pebbles, a 1/4 cup of soil, and 3 Tbsp of salt for each group. Have students choose which material best represents a stream's bed load (the pebbles), suspended load (the soil), and dissolved load (the salt).

Have students add all three materials to the jar and then shake it carefully to simulate a stream's load. After the contents have been thoroughly mixed, ask students to hypothesize how they could remove each material from the jar. (The pebbles settle to the bottom and can be picked out easily. If the water remains still long enough, the sediment will settle to the bottom of the jar. The salt can be removed if the water is evaporated.) LS **Kinesthetic**

load the materials carried by a stream

Load

The materials carried by a stream are called the stream's **load**. The size of a stream's load is affected by the stream's speed. Fast-moving streams can carry large particles. Rocks and pebbles bounce and scrape along the bottom and sides of the stream bed. Thus, the size of a stream's load also affects its rate of erosion. The illustration below shows the three ways that a stream can carry its load.

A stream can bounce large materials, such as pebbles and boulders, along the stream bed. These rocks are called the **bed load.**

A stream can carry small rocks and soil in suspension. These materials, called the **suspended load,** make the river look muddy.

The **dissolved load** is material carried in solution, which means that the material is dissolved in the water. Sodium and calcium are some of the materials in the dissolved load.

ACTIVITY — ADVANCED

River Field Guide Have students make a field guide for rivers. Suggest that they include photos of streams and rivers and write captions to describe each photo. The captions should incorporate terms such as *gradient, erosion, load, channel,* and *meanders.* Finally, have students hypothesize the river's speed, load, and erosional capacity. LS **Visual**

English Language Learners

CONNECTION to Geography — GENERAL

Amazon Tours Web Page The Amazon River basin is the world's largest watershed. It has an area of about 6 million square kilometers, which is almost twice as big as the Mississippi River watershed! Have students work in groups to design a Web page that describes the people, plants, and animals of the Amazon River basin. LS **Verbal/Visual**

The Stages of a River

In the early 1900s, William Morris Davis developed a model for the stages of river development. According to his model, rivers evolve from a youthful stage to an old-age stage. He thought that all rivers erode in the same way and at the same rate.

Today, scientists support a different model that considers factors of stream development that differ from those considered in Davis's model. For example, because different materials erode at different rates, one river may develop more quickly than another river. Many factors, including climate, gradient, and load, influence the development of a river. Scientists no longer use Davis's model to explain river development, but they still use many of his terms to describe a river. These terms describe a river's general features, not a river's actual age.

Youthful Rivers

A youthful river, such as the one shown in **Figure 5,** erodes its channel deeper rather than wider. The river flows quickly because of its steep gradient. Its channel is narrow and straight. The river tumbles over rocks in rapids and waterfalls. Youthful rivers have very few tributaries.

Mature Rivers

A mature river, as shown in **Figure 6,** erodes its channel wider rather than deeper. The gradient of a mature river is not as steep as that of a youthful river. Also, a mature river has fewer falls and rapids. A mature river is fed by many tributaries. Because of its good drainage, a mature river has more discharge than a youthful river.

✓ Reading Check What are the characteristics of a mature river?

CONNECTION TO Language Arts

Huckleberry Finn Mark Twain's famous book, *The Adventures of Huckleberry Finn,* describes the life of a boy who lived on the Mississippi River. Mark Twain's real name was Samuel Clemens. Do research to find out why Clemens chose to use the name Mark Twain and how the name relates to the Mississippi River.

▲ **Figure 5** *This youthful river is located in Yellowstone National Park in Wyoming. Rapids and falls are found where the river flows over hard, resistant rock.*

◄ **Figure 6** *A mature river, such as this one in the Amazon basin of Peru, curves back and forth. The bends in the river's channel are called* meanders.

Quiz — GENERAL

1. How does rainfall or snowmelt affect a river's discharge and ability to cause erosion?
(Rainfall and snowmelt increase both a river's discharge and the amount of erosion the river can cause.)

2. Explain the three types of materials carried by a river.
(bed load: large materials that roll or bounce along a riverbed; suspended load: materials that float suspended in a river; dissolved load: materials dissolved in a river)

Alternative Assessment — GENERAL

Modeling River Features Display the teaching transparency entitled "Rivers," and discuss the labeled features. Then, divide students into groups, and have each group use modeling clay to make a model of a river. The model should incorporate the concepts they have learned in this chapter. Students can use pins with attached labels to indicate river features. **LS Kinesthetic**

Figure 7 *This old river is located in New Zealand.*

Old Rivers

An old river has a low gradient and little erosive energy. Instead of widening and deepening its banks, the river deposits rock and soil in and along its channel. Old rivers, such as the one in **Figure 7,** are characterized by wide, flat *flood plains*, or valleys, and many bends. Also, an old river has fewer tributaries than a mature river because the smaller tributaries have joined together.

Rejuvenated Rivers

Rejuvenated (ri JOO vuh NAYT ed) rivers are found where the land is raised by tectonic activity. When land rises, the river's gradient becomes steeper, and the river flows more quickly. The increased gradient of a rejuvenated river allows the river to cut more deeply into the valley floor. Steplike formations called *terraces* often form on both sides of a stream valley as a result of rejuvenation. Can you find the terraces in **Figure 8**?

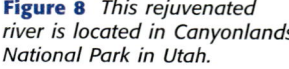 **Reading Check** How do rejuvenated rivers form?

Figure 8 *This rejuvenated river is located in Canyonlands National Park in Utah.*

Answer to Reading Check
Rejuvenated rivers form when the land is raised by tectonic forces.

BRAIN FOOD

Most river water comes from rainfall and melted snow that flow down from mountains. Ask students, "Why do some rivers continue to flow during a severe drought?"

(These rivers are probably lower than the water table, so their flow is maintained by seepage from groundwater.)

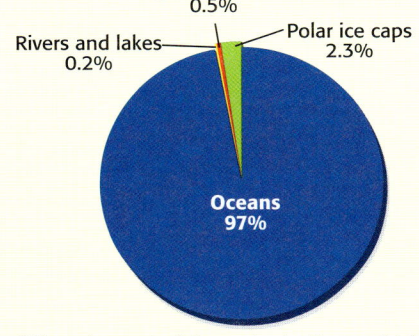

Summary

- Rivers cause erosion by removing and transporting soil and rock from the riverbed.
- The water cycle is the movement of Earth's water from the ocean to the atmosphere to the land and back to the ocean.
- A river system is made up of a network of streams and rivers.
- A watershed is a region that collects runoff water that then becomes part of a river or a lake.

- A stream with a high gradient has more energy for eroding soil and rock.
- When a stream's discharge increases, its erosive energy also increases.
- A stream with a load of large particles has a higher rate of erosion than a stream with a dissolved load.
- A developing river can be described as youthful, mature, old, or rejuvenated.

Using Key Terms

1. Use each of the following terms in a separate sentence: *erosion, water cycle, tributary, watershed, divide, channel,* and *load.*

Understanding Key Ideas

2. Which of the following drains a watershed?
   a. a divide
   b. a drainage basin
   c. a tributary
   d. a water system

3. Describe how the Grand Canyon was formed.

4. Draw the water cycle. In your drawing, label *condensation, precipitation,* and *evaporation.*

5. What are three factors that affect the rate of stream erosion?

6. Which stage of river development is characterized by flat flood plains?

Critical Thinking

7. **Making Inferences** How does the water cycle help develop river systems?

8. **Making Comparisons** How do youthful rivers, mature rivers, and old rivers differ?

Interpreting Graphics

Use the pie graph below to answer the questions that follow.

Distribution of Water in the World

Water underground, in soil, and in air
0.5%

Rivers and lakes
0.2%

Polar ice caps
2.3%

Oceans
97%

9. Where is most of the water in the world found?

10. In what form is the majority of the world's fresh water?

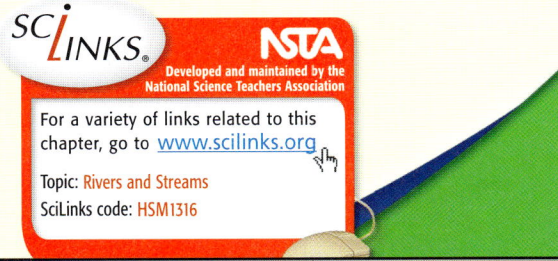

SCILINKS.

NSTA

Developed and maintained by the National Science Teachers Association

For a variety of links related to this chapter, go to www.scilinks.org

Topic: Rivers and Streams
SciLinks code: HSM1316

Answers to Section Review

1. Sample answer: The Grand Canyon was formed by erosion. Rain is a part of the water cycle. A tributary joins a larger river. A watershed can also be called a drainage basin. Watersheds are separated by a divide. A channel is a path that a stream follows. A river's load often includes rocks and pebbles.

2. d

3. The Grand Canyon was formed by erosion caused by the Colorado River washing away soil and rock from the riverbed.

4. Answers may vary. The drawing should resemble Figure 2.

5. gradient, discharge, and load

6. old rivers

7. Answers may vary. The water cycle causes continuous movement of water between the Earth's surface and the atmosphere. The flow of water to the oceans creates river systems.

8. Answers may vary. Youthful rivers erode deep channels. Mature rivers erode wide channels. Old rivers deposit sediment in their channels and along their banks. Rejuvenated rivers form terraces in the river valley.

9. Most of the world's water is found in the oceans.

10. The majority of fresh water is in the form of ice.

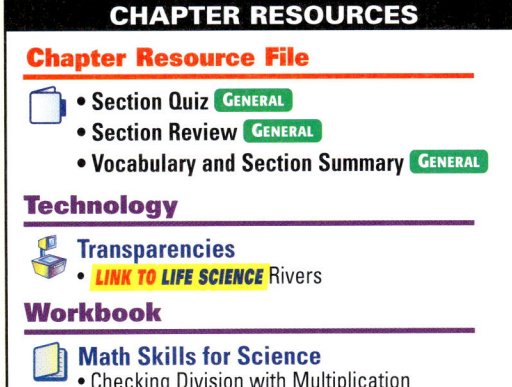

CHAPTER RESOURCES

Chapter Resource File

- Section Quiz GENERAL
- Section Review GENERAL
- Vocabulary and Section Summary GENERAL

Technology

- Transparencies
  - **LINK TO LIFE SCIENCE** Rivers

Workbook

- Math Skills for Science
  - Checking Division with Multiplication

SECTION
2

Focus

Overview

This section describes the variety of ways that the load carried by a river can be deposited. Students will learn about the different landforms created by river deposits, and they will explore the connections between river deposits, floods, and agriculture.

Bellringer

Post the following question on the board or overhead projector: "Even though flooding along rivers is potentially harmful, many farms are located near rivers. Why do people build farms along rivers?" (The flood waters deposit sediments that contribute to the land's fertility.)

Motivate

Demonstration — GENERAL

Modeling Deposition To show students how sediment tends to sort according to particle size, place a handful of soil, sand, and small gravel in a large plastic jar filled with water. Shake the jar vigorously, and then place it on your desk. Ask students to describe what happens to the different-sized particles in the jar as time passes. **LS Visual**

READING WARM-UP

Objectives

● Describe the four different types of stream deposits.

● Describe how the deposition of sediment affects the land.

Terms to Learn

deposition   alluvial fan
delta   floodplain

READING STRATEGY

Prediction Guide Before reading this section, write the title of each heading in this section. Next, under each heading, write what you think you will learn.

Stream and River Deposits

If your job were to carry millions of tons of soil across the United States, how would you do it? You might use a bulldozer or a dump truck, but it would still take you a long time. Did you know that rivers do this job every day?

Rivers erode and move enormous amounts of material, such as soil and rock. Acting as liquid conveyor belts, rivers often carry fertile soil to farmland and wetlands. Although erosion is a serious problem, rivers also renew soils and form new land. As you will see in this section, rivers create some of the most impressive landforms on Earth.

Deposition in Water

You have learned how flowing water erodes the Earth's surface. After rivers erode rock and soil, they drop, or *deposit,* their load downstream. **Deposition** is the process in which material is laid down or dropped. Rock and soil deposited by streams are called *sediment.* Rivers and streams deposit sediment where the speed of the water current decreases. **Figure 1** shows this type of deposition.

Figure 1 *This photo shows erosion and deposition at a bend, or meander, of a river in Alaska.*

Deposition occurs along the inside bank of the bend, where the water flows slower.

Erosion occurs along the outside bank of the bend, where the water flows faster.

CHAPTER RESOURCES

Chapter Resource File

- Lesson Plan
- Directed Reading A **BASIC**
- Directed Reading B **SPECIAL NEEDS**

Technology

 **Transparencies**
- Bellringer

WEIRD SCIENCE

The Okavango River, in Africa, flows approximately 1,600 km through Angola, Namibia, and Botswana before it empties into the middle of the Kalahari Desert, where the river evaporates! The river's alluvial fan provides a haven for the desert's plants and animals.

Placer Deposits

Heavy minerals are sometimes deposited at places in a river where the current slows down. This kind of sediment is called a *placer deposit* (PLAS uhr dee PAHZ it). Some placer deposits contain gold. During the California gold rush, which began in 1849, many miners panned for gold in the placer deposits of rivers, as shown in **Figure 2.**

Delta

A river's current slows when a river empties into a large body of water, such as a lake or an ocean. As its current slows, a river often deposits its load in a fan-shaped pattern called a **delta.** In **Figure 3,** you can see an astronaut's view of the Nile Delta. A delta usually forms on a flat surface and is made mostly of mud. These mud deposits form new land and cause the coastline to grow. The world's deltas are home to a rich diversity of plant and animal life.

If you look back at the map of the Mississippi River watershed, you can see where the Mississippi Delta has formed. It has formed where the Mississippi River flows into the Gulf of Mexico. Each of the fine mud particles in the delta began its journey far upstream. Parts of Louisiana are made up of particles that were transported from places as far away as Montana, Minnesota, Ohio, and Illinois!

✔ **Reading Check** What are deltas made of? (*See the Appendix for answers to Reading Checks.*)

Figure 2 *Miners rushed to California in the 1850s to find gold. They often found it in the bends of rivers in placer deposits.*

deposition the process in which material is laid down

delta a fan-shaped mass of material deposited at the mouth of a stream

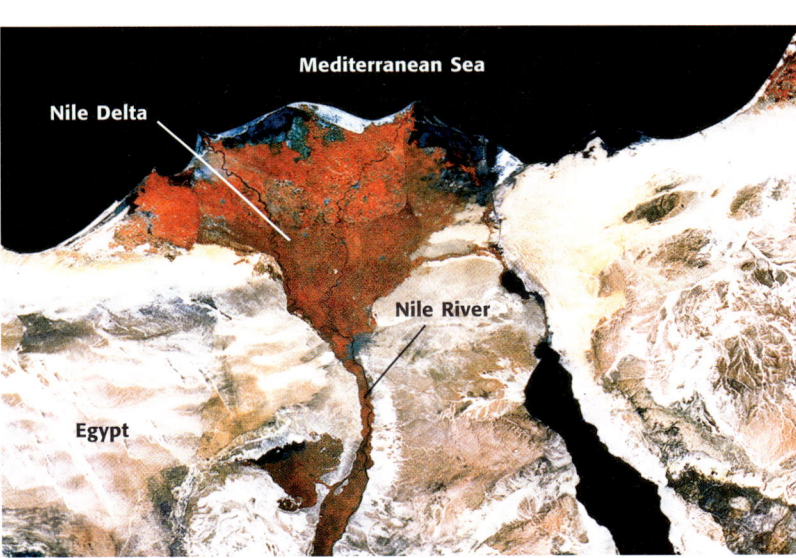

Mediterranean Sea

Nile Delta

Egypt

Nile River

Figure 3 *As sediment is dropped at the mouth of the Nile River, in Egypt, a delta forms.*

Answer to Reading Check
Deltas are made of the deposited load of the river, which is mostly mud.

Is That a Fact!

Under average conditions, the Mississippi River carries about 17,000 m³ of water by a given point every second. A small carry-on suitcase is about 0.03 m³, so watching the river go by on an average day is equivalent to watching about 566,666 water-filled suitcases pass by you every second!

Reteaching — BASIC

Prospecting Using a large map of a local river, have students label areas that are likely to have placer deposits or flood plains. Also, have students identify areas where the river's channel might be likely to shift over time. LS **Visual**

Quiz — GENERAL

1. A river runs down a rapids, eases through a valley for about 3 km, and then tumbles down a waterfall into a lake. Where along this path would you most likely find a placer deposit? Why? (in the valley, because heavy minerals are often deposited where currents are slow)

2. Define flood plain. (land that is periodically flooded when a river overflows its banks)

Alternative Assessment — GENERAL

Where the River Ends Have students use modeling clay to create a model of a delta and an alluvial fan. They should also model the immediate environment around each river feature and use pins with labels to identify the processes involved in the formation of each feature. LS **Kinesthetic**

Figure 4 An alluvial fan, like this one at Death Valley in California, forms when an eroding stream changes rapidly into a depositing stream.

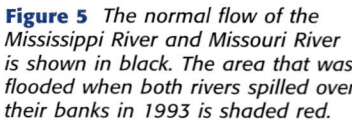

alluvial fan a fan-shaped mass of material deposited by a stream when the slope of the land decreases sharply

floodplain an area along a river that forms from sediments deposited when the river overflows its banks

Deposition on Land

When a fast-moving mountain stream flows onto a flat plain, the stream slows down very quickly. As the stream slows down, it deposits sediment. The sediment forms an alluvial fan, such as the one shown in **Figure 4. Alluvial fans** are fan-shaped deposits that, unlike deltas, form on dry land.

Floodplains

During periods of high rainfall or rapid snow melt, a sudden increase in the volume of water flowing into a stream can cause the stream to overflow its banks. The area along a river that forms from sediment deposited when a river overflows its banks is called a **floodplain.** When a stream floods, a layer of sediment is deposited across the flood plain. Each flood adds another layer of sediment.

Flood plains are rich farming areas because periodic flooding brings new soil to the land. However, flooding can cause damage, too. When the Mississippi River flooded in 1993, farms were destroyed, and entire towns were evacuated. **Figure 5** shows an area north of St. Louis, Missouri, that was flooded.

Figure 5 The normal flow of the Mississippi River and Missouri River is shown in black. The area that was flooded when both rivers spilled over their banks in 1993 is shaded red.

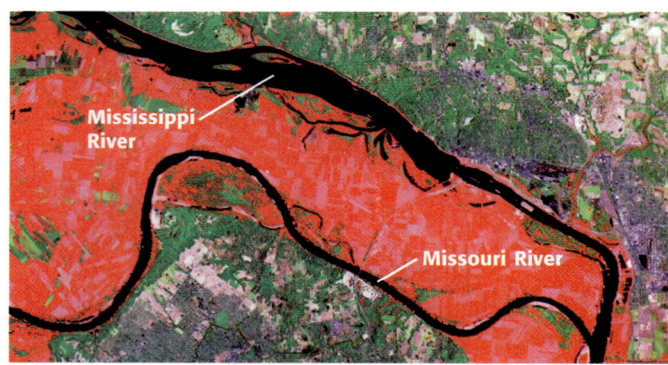

SCIENCE HUMOR

Q: What is a flood?

A: It's a river that's too big for its bridges.

Flooding Dangers

The flooding of the Mississippi River in 1993 caused damage in nine states. But floods can damage more than property. Many people have lost their lives to powerful floods. As shown in **Figure 6,** flash flooding can take a driver by surprise. However, there are ways that floods can be controlled.

One type of barrier that can be built to help control flooding is called a *dam*. A dam is a barrier that can redirect the flow of water. A dam can prevent flooding in one area and create an artificial lake in another area. The water stored in the artificial lake can be used to irrigate farmland during droughts and provide drinking water to local towns and cities. The stored water can also be used to generate electricity.

Overflow from a river can also be controlled by a barrier called a *levee*. A levee is the buildup of sediment deposited along the channel of a river. This buildup helps keep the river inside its banks. People often use sandbags to build artificial levees to control water during serious flooding.

✓ Reading Check List two ways that the flow of water can be controlled.

Figure 6 *Cars driven on flooded roads can easily be carried down to deeper, more dangerous water.*

SECTION Review

Summary

- Sediment forms several types of deposits.
- Sediments deposited where a river's current slows are called *placer deposits.*
- A delta is a fan-shaped deposit of sediment where a river meets a large body of water.
- Alluvial fans can form when a river deposits sediment on land.
- Flooding brings rich soil to farmland but can also lead to property damage and death.

Using Key Terms

1. In your own words, write a definition for each of the following terms: *deposition* and *flood plain.*

Understanding Key Ideas

2. Which of the following forms at places in a river where the current slows?
   a. a placer deposit
   b. a delta
   c. a flood plain
   d. a levee

3. Which of the following can help prevent a flood?
   a. a placer deposit
   b. a delta
   c. a flood plain
   d. a levee

4. Where do alluvial fans form?

5. Explain why flood plains are both good and bad areas for farming.

Math Skills

6. A river flows at a speed of 8 km/h. If you floated on a raft in this river, how far would you have traveled after 5 h?

Critical Thinking

7. **Identifying Relationships** What factors increase the likelihood that sediment will be deposited?

8. **Making Comparisons** How are alluvial fans and deltas similar?

SCILINKS.
Developed and maintained by the National Science Teachers Association

For a variety of links related to this chapter, go to www.scilinks.org

Topic: Stream Deposits
SciLinks code: HSM1458

Answers to Section Review

1. Sample answer: Deposition is material being laid down or dropped. Flood plains are areas where sediment is deposited when a river overflows its banks.
2. a
3. d
4. Alluvial fans form where a fast-moving mountain stream flows onto a flat plain and deposits sediment in a fan-shaped pattern.
5. Answers may vary. Flood plains form fertile farmland because flood waters periodically deposit new soil on the land. However, flooding can damage crops.
6. 8 km/h × 5 h = 40 km
7. Answers may vary. Factors that reduce a river's speed increase the chance that a river's load will be deposited.
8. Answers may vary. Both deltas and alluvial fans are fan-shaped deposits formed when a stream slows. Deltas form as rivers enter larger bodies of water, and alluvial fans form on dry land.

Answer to Reading Check

The flow of water can be controlled by dams and levees.

CHAPTER RESOURCES

Chapter Resource File

- Section Quiz **GENERAL**
- Section Review **GENERAL**
- Vocabulary and Section Summary **GENERAL**
- Reinforcement Worksheet **BASIC**

Overview

In this section, students will learn about groundwater. The section discusses the formation of aquifers and how surface water enters them. Students will learn how wells and springs bring groundwater to the surface. Finally, the section discusses how the movement of groundwater forms caves.

Bellringer

Ask students "A family lives 50 km from the nearest stream or lake and gets water from a well. Where does the water in the well come from?" (It comes from water stored underground.)

Motivate

Demonstration — GENERAL

Groundwater Model Layer an aquarium half full with gravel, sand, and potting soil. Pack the material firmly. Add water until you can clearly see areas of aeration and saturation. Ask students to compare this model to **Figure 1.** Using a marker, draw a line for the water table on the glass. Introduce the terms *zone of saturation* and *zone of aeration.* Discuss the difference between the terms. **LS Visual**

SECTION
3

READING WARM-UP

Objectives

- Identify and describe the location of the water table.
- Describe an aquifer.
- Explain the difference between a spring and a well.
- Explain how caves and sinkholes form as a result of erosion and deposition.

Terms to Learn

water table
aquifer
porosity
permeability
recharge zone
artesian spring

READING STRATEGY

Discussion Read this section silently. Write down questions that you have about this section. Discuss your questions in a small group.

water table the upper surface of underground water; the upper boundary of the zone of saturation

Figure 1 *The water table is the upper surface of the zone of saturation.*

Water Underground

Imagine that instead of turning on a faucet to get a glass of water, you pour water from a chunk of solid rock! This idea may sound crazy, but millions of people get their water from within rock that is deep underground.

Although you can see some of Earth's water in streams and lakes, you cannot see the large amount of water that flows underground. The water located within the rocks below the Earth's surface is called *groundwater.* Groundwater not only is an important resource but also plays an important role in erosion and deposition.

The Location of Groundwater

Surface water seeps underground into the soil and rock. This underground area is divided into two zones. Rainwater passes through the upper zone, called the *zone of aeration.* Farther down, the water collects in an area called the *zone of saturation.* In this zone, the spaces between the rock particles are filled with water.

These two zones meet at a boundary known as the **water table,** shown in **Figure 1.** The water table rises during wet seasons and falls during dry seasons. In wet regions, the water table can be at or just beneath the soil's surface. In dry regions, such as deserts, the water table may be hundreds of meters beneath the ground.

✓ **Reading Check** Describe where the zone of aeration is located. *(See the Appendix for answers to Reading Checks.)*

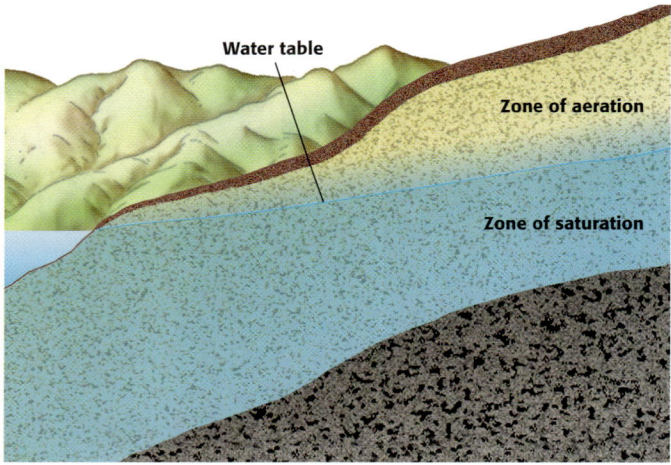

CHAPTER RESOURCES

Chapter Resource File

- **Lesson Plan**
- **Directed Reading A** BASIC
- **Directed Reading B** SPECIAL NEEDS

Technology

 Transparencies
- Bellringer
- The Water Table and Wells

Answer to Reading Check

The zone of aeration is located underground. It is the area above the water table.

Aquifers

A rock layer that stores groundwater and allows the flow of groundwater is called an **aquifer.** An aquifer can be described by its ability to hold water and its ability to allow water to pass freely through it.

Porosity

The more open spaces, or pores, between particles in an aquifer, the more water the aquifer can hold. The percentage of open space between individual rock particles in a rock layer is called **porosity.**

Porosity is influenced by the differences in sizes of the particles in the rock layer. If a rock layer contains many particles of different sizes, it is likely that small particles will fill up the different-sized empty spaces between large particles. Therefore, a rock layer with particles of different sizes has a low percentage of open space between particles and has low porosity. On the other hand, a rock layer containing same-sized particles has high porosity. This rock layer has high porosity because smaller particles are not present to fill the empty space between particles. So, there is more open space between particles.

Permeability

If the pores of a rock layer are connected, groundwater can flow through the rock layer. A rock's ability to let water pass through is called **permeability.** A rock that stops the flow of water is *impermeable*.

The larger the particles are, the more permeable the rock layer is. Because large particles have less surface area relative to their volume than small particles do, large particles cause less friction. *Friction* is a force that causes moving objects to slow down. Less friction allows water to flow more easily through the rock layer, as shown in **Figure 2.**

aquifer a body of rock or sediment that stores groundwater and allows the flow of groundwater

porosity the percentage of the total volume of a rock or sediment that consists of open spaces

permeability the ability of a rock or sediment to let fluids pass through its open spaces, or pores

INTERNET ACTIVITY

For another activity related to this chapter, go to **go.hrw.com** and type in the keyword **HZ5DEPW.**

Figure 2 *Large particles, shown at left, have less total surface area—and so cause less friction— than small particles, shown at right, do.*

BRAIN FOOD

The Great Artesian Basin
Much of Australia's groundwater is stored in artesian formations that are fed by a vast underground aquifer called the Great Artesian Basin. Unfortunately, most of the water is too salty for people to drink or use for irrigation. Have students find out how this water could be processed in a desalination plant.

Answer to Reading Check
The size of the recharge zone depends on how permeable rock is at the surface.

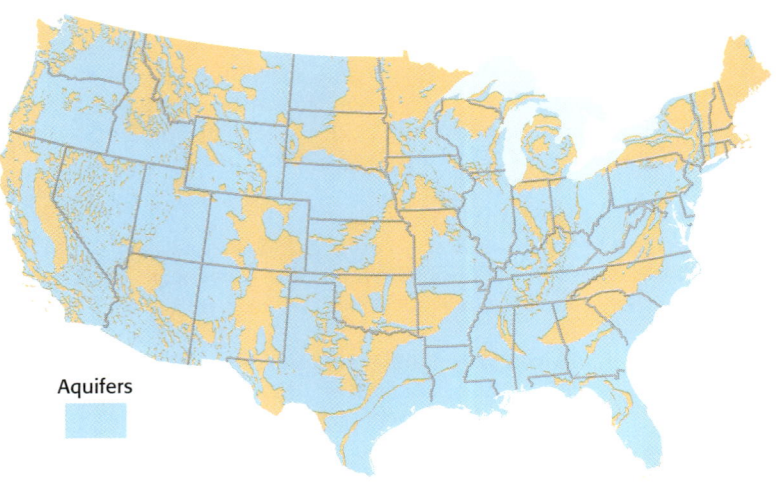

Figure 3 *This map shows aquifers in the United States (excluding Alaska and Hawaii).*

Aquifers

recharge zone an area in which water travels downward to become part of an aquifer

Water Conservation
Did you know that water use in the United States has been reduced by 15% in the last 20 years? This decrease is due in part to the conservation efforts of people like you. Work with a parent to create a water budget for your household. Figure out how much water your family uses every day. Identify ways to reduce your water use, and then set a goal to limit your water use over the course of a week.

Aquifer Geology and Geography

The best aquifers usually form in permeable materials, such as sandstone, limestone, or layers of sand and gravel. Some aquifers cover large underground areas and are an important source of water for cities and agriculture. The map in **Figure 3** shows the location of the major aquifers in the United States.

Recharge Zones

Like rivers, aquifers depend on the water cycle to maintain a constant flow of water. The ground surface where water enters an aquifer is called the **recharge zone.** The size of the recharge zone depends on how permeable rock is at the surface. If the surface rock is permeable, water can seep down into the aquifer. If the aquifer is covered by an impermeable rock layer, water cannot reach the aquifer. Construction of buildings on top of the recharge zone can also limit the amount of water that enters an aquifer.

✔ **Reading Check** What factors affect the size of the recharge zone?

Springs and Wells

Groundwater movement is determined by the slope of the water table. Like surface water, groundwater tends to move downslope, toward lower elevations. If the water table reaches the Earth's surface, water will flow out from the ground and will form a *spring*. Springs are an important source of drinking water. In areas where the water table is higher than the Earth's surface, lakes will form.

INCLUSION Strategies

• **Attention Deficit Disorder**   • **Learning Disabled**
• **English Language Learners**

Have students play a card game to help them learn the vocabulary in this section. Organize students into groups of three. Have one student in the group write one vocabulary term on each card. Have another student make cards with one definition per card. Have the third student in the group draw an illustration for each vocabulary term on a card. When students are finished, have them exchange their cards with another group and try to match the term with the definition and illustration. **English Language Learners**

LS Verbal/Visual Co-op Learning

Artesian Springs

A sloping layer of permeable rock sandwiched between two layers of impermeable rock is called an *artesian formation*. The permeable rock is an aquifer, and the top layer of impermeable rock is called a *cap rock*, as shown in **Figure 4**. Artesian formations are the source of water for artesian springs. An **artesian spring** is a spring whose water flows from a crack in the cap rock of the aquifer. Artesian springs are sometimes found in deserts, where they are often the only source of water.

Most springs have cool water. However, some springs have hot water. The water becomes hot when it flows deep in the Earth, because Earth's temperature increases with depth. The temperature of some hot springs can reach 50°C!

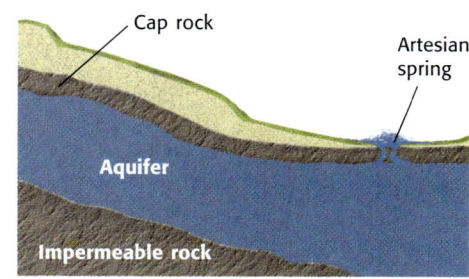

Figure 4 *Artesian springs form when water from an aquifer flows through cracks in the cap rock of an artesian formation.*

Wells

A human-made hole that is deeper than the level of the water table is called a *well*. If a well is not deep enough, as shown in **Figure 5,** it will dry up when the water table falls below the bottom of the well. Also, if an area has too many wells, groundwater can be removed too rapidly. If groundwater is removed too rapidly, the water table will drop, and all of the wells will run dry.

artesian spring a spring whose water flows from a crack in the cap rock over the aquifer

✓ **Reading Check** How deep must a well be to reach water?

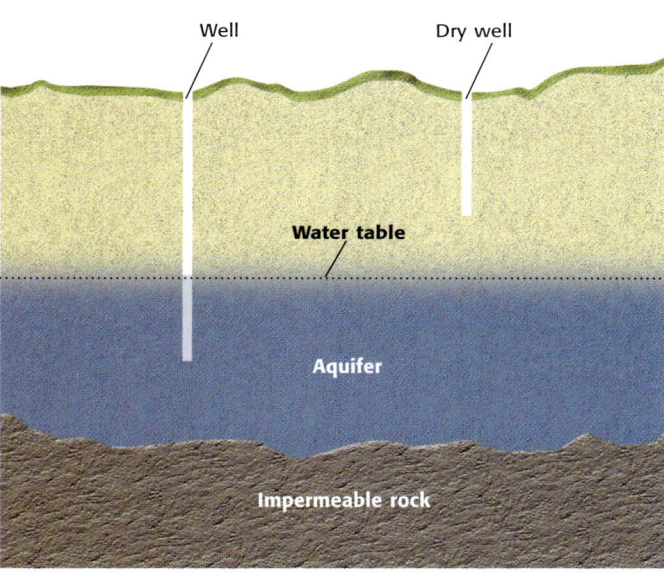

Figure 5 *A well must be drilled deep enough so that when the water table drops, the well still contains water.*

CONNECTION to Life Science ── GENERAL

Life in the Desert Elephants that live in the desert of Namibia may travel four days to find a water hole. Once there, they might have to dig a meter into the sand to reach water. Antelopes, such as springbok and eland, and other desert animals depend on the elephants to penetrate the water table. Have students research ways in which desert plants and animals have adapted to regions that have little surface water. **LS** Verbal

Using the Figure ── ADVANCED

Wells and Artesian Springs Guide students through **Figure 4.** List the labels on the board, and help students define each term. Explain that artesian springs are driven by hydraulic pressure. Artesian springs are found in areas where the water table is above the outlet for the spring. If students have ever siphoned a liquid, they have applied the principle that causes artesian springs to flow. Challenge students to demonstrate an artesian spring using a water-filled container and a length of tubing. **LS** Visual — English Language Learners

Answer to Reading Check

A well must be deeper than the water table for it to be able to reach water.

Reteaching — BASIC

Spring Model Give students a large piece of floral foam and a plastic container with a lid. Have students soak the foam with water and place it in the container. Tell students to cut a small hole in the plastic lid and cut the edge off the lid so that it fits just inside the container. Then, have students press the lid down onto the foam. Ask students, "What happens?" (Water is forced up through the hole by the pressure from the lid.) **LS Visual/Kinesthetic**

Quiz — GENERAL

1. Define porosity and permeability. (Porosity is the amount of space between rock particles. Permeability is the ability of rock to allow water to flow.)

2. What is a spring? What is a well? (A spring forms where the water table reaches the surface and water flows out. A well is a human construction that extends below the water table.)

Alternative Assessment — GENERAL

 Cave Tours Have students research a cave system in the United States. Then, help students construct a virtual cave. Have them create models of the cave's formations using modeling clay. Students can also make an audio tape tour of the cave that tells visitors about the cave's features and how the features formed. **LS Kinesthetic**

CONNECTION TO Environmental Science

Bat Environmentalists
Most bat species live in caves. Bats are night-flying mammals that play an important role in the environment. Bats eat vast quantities of insects. Many bat species also pollinate plants and distribute seeds. Can you think of other animals that eat insects, pollinate plants, and distribute seeds? Create a poster that includes pictures of these other animals. **ACTIVITY**

Figure 6 *At Carlsbad Caverns in New Mexico, underground passages and enormous "rooms" have been eroded below the surface of the Earth.*

Underground Erosion and Deposition

As you have learned, rivers cause erosion when water removes and transports rock and soil from its banks. Groundwater can also cause erosion. However, groundwater causes erosion by dissolving rock. Some groundwater contains weak acids, such as carbonic acid, that dissolve the rock. Also, some types of rock, such as limestone, dissolve in groundwater more easily than other types do.

When underground erosion happens, caves can form. Most of the world's caves formed over thousands of years as groundwater dissolved the limestone of the cave sites. Some caves, such as the one shown in **Figure 6**, reach spectacular proportions.

Cave Formations

Although caves are formed by erosion, they also show signs of deposition. Water that drips from a crack in a cave's ceiling leaves behind deposits of calcium carbonate. Sharp, icicle-shaped features that form on cave ceilings are known as *stalactites* (stuh LAK tiets). Water that falls to the cave's floor adds to cone-shaped features known as *stalagmites* (stuh LAG miets). If water drips long enough, the stalactites and stalagmites join to form a *dripstone column.*

✓ Reading Check What process causes the formation of stalactites and stalagmites?

Stalactite

Stalagmite

Answer to Reading Check
Deposition is the process that causes the formation of stalactites and stalagmites.

Sinkholes

When the water table is lower than the level of a cave, the cave is no longer supported by the water underneath. The roof of the cave can then collapse, which leaves a circular depression called a *sinkhole*. Surface streams can "disappear" into sinkholes and then flow through underground caves. Sinkholes often form lakes in areas where the water table is high. Central Florida is covered with hundreds of round sinkhole lakes. **Figure 7** shows how the collapse of an underground cave can affect a landscape.

Figure 7 *The damage to this city block shows the effects of a sinkhole in Winter Park, Florida.*

SECTION Review

Summary

- The water table is the boundary between the zone of aeration and the zone of saturation.
- Porosity and permeability describe an aquifer's ability to hold water and ability to allow water to flow through.
- Springs are a natural way that water reaches the surface. Wells are made by humans.
- Caves and sinkholes form from the erosion of limestone by groundwater.

Using Key Terms

1. Use the following terms in the same sentence: *water table, aquifer, porosity,* and *artesian spring*.

Understanding Key Ideas

2. Which of the following describes an aquifer's ability to allow water to flow through?
   a. porosity
   b. permeability
   c. geology
   d. recharge zone

3. What is the water table?

4. Describe how particles affect the porosity of an aquifer.

5. Explain the difference between an artesian spring and other springs.

6. Name a feature that is formed by underground erosion.

7. Name two features that are formed by underground deposition.

8. What type of weathering process causes underground erosion?

Math Skills

9. Groundwater in an area flows at a speed of 4 km/h. How long would it take the water to flow 10 km to its spring?

Critical Thinking

10. **Predicting Consequences** Explain how urban growth might affect the recharge zone of an aquifer.

11. **Making Comparisons** Explain the difference between a spring and a well.

12. **Analyzing Relationships** What is the relationship between the zone of aeration, the zone of saturation, and the water table?

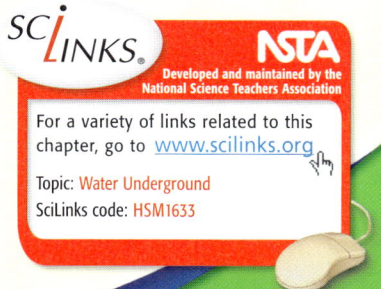

SCLINKS.

NSTA
Developed and maintained by the
National Science Teachers Association

For a variety of links related to this chapter, go to www.scilinks.org

Topic: Water Underground
SciLinks code: HSM1633

Answers to Section Review

1. Sample answer: Depending on the porosity of the ground and the level of the water table, groundwater can exist in formations such as aquifers or artesian springs.

2. b

3. A water table is the upper surface of the zone of saturation.

4. Answers may vary. A rock layer that contains different-sized particles has less open space between particles and is less porous than a rock layer that contains same-sized particles.

5. Answers may vary. An artesian spring occurs if water flows upward through the cap rock and flows out at the surface. Other springs do not have a layer of cap rock.

6. Caves form by underground erosion.

7. Stalagmites, stalactites, and dripstone columns form by underground deposition.

8. Groundwater causes erosion by dissolving rock.

9. $10 \text{ km} \div 4 \text{ km/h} = 2.5 \text{ h}$

10. Answers may vary. Surface water in a watershed enters the ground in the recharge zone. A parking lot is an impermeable surface layer, so it could reduce or prevent the flow of water into the recharge zone.

11. Answers may vary. A spring is a place where the water table reaches the surface. A well is a human-made hole dug deeper than the level of the water table.

12. Sample answer: The zone of aeration is above the water table. The zone of saturation is below the water table, and the spaces between the rocks are filled with water.

Overview

This section discusses water pollution, overuse, and treatment. Students will learn about the difference between point-source and nonpoint-source pollution and explore different methods for treating polluted water. The section discusses trends in domestic, industrial, and agricultural water use and conservation.

 **Bellringer**

Write the following scenario on the board or overhead projector: "While hiking, you realize your canteen is almost empty." Then ask, "Why should you not fill the canteen with water from the nearby stream?" (Even though the water may look clean, it might contain pollutants or bacteria.)

Discussion —— GENERAL

Watching Water Use Ask students to list ways that people use water. Write the responses on the board. Ask students to track the amount of water they use during the week. Have them use this number to estimate how much water they will use in their lifetime. **LS Verbal**

READING WARM-UP

Objectives

- Identify two forms of water pollution.
- Explain how the properties of water influence the health of a water system.
- Describe two ways that wastewater can be treated.
- Describe how water is used and how water can be conserved in industry, in agriculture, and at home.

Terms to Learn

point-source pollution
nonpoint-source pollution
sewage treatment plant
septic tank

READING STRATEGY

Paired Summarizing Read this section silently. In pairs, take turns summarizing the material. Stop to discuss ideas that seem confusing.

point-source pollution pollution that comes from a specific site

nonpoint-source pollution pollution that comes from many sources rather than from a single, specific site

Figure 1 The runoff from this irrigation system could collect pesticides and other pollutants. The result would be nonpoint-source pollution.

Using Water Wisely

Did you know that you are almost 65% water? You depend on clean, fresh drinking water to maintain that 65% of you. But there is a limited amount of fresh water available on Earth. Only 3% of Earth's water is drinkable.

And of the 3% of Earth's water that is drinkable, 75% is frozen in the polar icecaps. This frozen water is not readily available for our use. Therefore, it is important that we protect our water resources.

Water Pollution

Surface water, such as the water in rivers and lakes, and groundwater can be polluted by waste from cities, factories, and farms. Pollution is the introduction of harmful substances into the environment. Water can become so polluted that it can no longer be used or can even be deadly.

Point-Source and Nonpoint-Source Pollution

Pollution that comes from one specific site is called **point-source pollution.** For example, a leak from a sewer pipe is point-source pollution. In most cases, this type of pollution can be controlled because its source can be identified.

Nonpoint-source pollution, another type of pollution, is pollution that comes from many sources. This type of pollution is much more difficult to control because it does not come from a single source. Most nonpoint-source pollution reaches bodies of water by runoff. The main sources of nonpoint-source pollution are street gutters, fertilizers, eroded soils and silt from farming and logging, drainage from mines, and salts from irrigation. **Figure 1** shows an example of a source of nonpoint-source pollution.

✔ **Reading Check** What type of pollution is the hardest to control? (*See the Appendix for answers to Reading Checks.*)

CHAPTER RESOURCES

Chapter Resource File

- Lesson Plan
- Directed Reading A BASIC
- Directed Reading B SPECIAL NEEDS

Technology

 Transparencies
- Bellringer

Answer to Reading Check
Nonpoint-source pollution is hardest to control.

Figure 2 *Waste from farm animals can seep into groundwater and cause nitrate pollution.*

Health of a Water System

You might not realize it, but water quality affects your quality of life as well as other organisms that depend on water. Therefore, it is important to understand how the properties of water influence water quality.

Dissolved Oxygen

Just as you need oxygen to live, so do fish and other organisms that live in lakes and streams. The oxygen dissolved in water is called *dissolved oxygen,* or DO. Levels of DO that are below 4.0 mg/L in fresh water can cause stress and possibly death for organisms in the water.

Pollutants such as sewage, fertilizer runoff, and animal waste can decrease DO levels. Temperature changes also affect DO levels. For example, cold water holds more oxygen than warm water does. Facilities such as nuclear power plants can increase the temperature of lakes and rivers when they use the water as a cooling agent. Such an increase in water temperature is called *thermal pollution,* which causes a decrease in DO levels.

Nitrates

Nitrates are naturally occurring compounds of nitrogen and oxygen. Small amounts of nitrates in water are normal. However, elevated nitrate levels in water can be harmful to organisms. An excess of nitrates in lakes and rivers can also lower DO levels. As shown in **Figure 2,** nitrate pollution can come from animal wastes or fertilizers that seep into groundwater.

Alkalinity

Alkalinity refers to the water's ability to neutralize acid. Acid rain and other acid wastes can harm aquatic life. A pH below 6.0 is too acidic for most aquatic life. Water with a higher alkalinity can better protect organisms from acid.

Quick Lab

Measuring Alkalinity

1. Identify two water sources from which to collect water samples.
2. Fill a **plastic cup** with water from one source. Fill a **second plastic cup** with water from the second source. Label each cup with its source.
3. Using a **pH test kit,** test the pH of each sample.
4. Follow the instructions in the test kit, and determine the pH of each of the two samples. Record your observations.
5. What did the results for the two samples indicate about the two sources?
6. Use **water test kits** to measure DO and nitrate levels in the two water samples, and discuss your results.

Putting Pollution in its Place

Organize the class into small groups, and challenge each group to create a board game. Tell students that the object of the game is to be the first player to successfully travel through a sewage treatment plant. Provide each group with poster board, plain index cards, and markers. Direct them to create a game board that leads players through the plant and incorporates the concepts they have learned in this section. On the index cards, have them write clues and questions directing players' movements through the sewage treatment process. For example, they might write, "If you can define primary treatment, advance to the aeration tank." Have students create written rules, and allow time for the game to be played. **LS Visual/Kinesthetic**

Cleaning Polluted Water

When you flush the toilet or watch water go down the shower drain, do you ever wonder where the water goes? If you live in a city or large town, the water flows through sewer pipes to a sewage treatment plant. **Sewage treatment plants** are facilities that clean the waste materials out of water. These plants help protect the environment from water pollution. They also protect us from diseases that are easily transmitted through dirty water.

sewage treatment plant a facility that cleans the waste materials found in water that comes from sewers or drains

Primary Treatment

When water reaches a sewage treatment plant, it is cleaned in two ways. First, it goes through a series of steps known as *primary treatment*. In primary treatment, dirty water is passed through a large screen to catch solid objects, such as paper, rags, and bottle caps. The water is then placed in a large tank, where smaller particles, or sludge, can sink and be filtered out. These particles include things such as food, coffee grounds, and soil. Any floating oils and scum are skimmed off the surface.

Secondary Treatment

After undergoing primary treatment, the water is ready for *secondary treatment*. In secondary treatment, the water is sent to an aeration tank, where it is mixed with oxygen and bacteria. The bacteria feed on the wastes and use the oxygen. The water is then sent to another settling tank, where chlorine is added to disinfect the water. The water is finally released into a water source—a river, a lake, or the ocean. **Figure 3** shows the major components of a sewage treatment plant.

Figure 3 *If you live in a city, the water used in your home most likely ends up at a sewage treatment plant, where the water is cleaned.*

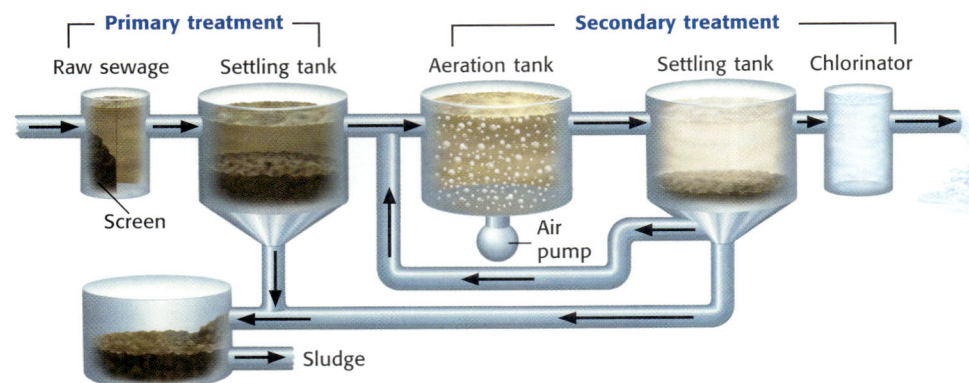

Primary treatment — Raw sewage, Settling tank, Screen

Secondary treatment — Aeration tank, Settling tank, Chlorinator, Air pump

Sludge

Waterborne Disease in North America
Most people in the United States and many students may think that waterborne diseases are a problem only in developing countries. But despite elaborate wastewater purification systems, outbreaks still occur. In 1993, the same year in which the *Cryptosporidium* outbreak affected 400,000 people in Milwaukee, a similar outbreak occurred in Round Rock, Texas. In 1993 and 1994, an epidemic of another, deadlier waterborne illness—cholera—swept through Latin America as far north as northern Mexico. The epidemic raised fears that the United States could eventually be affected.

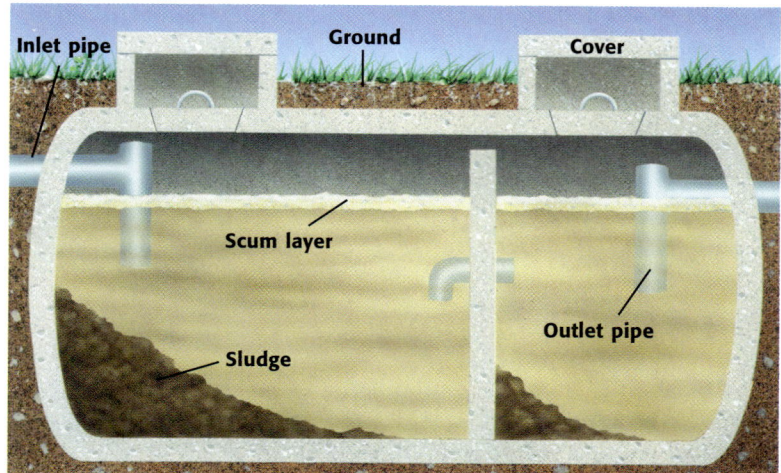

Figure 4 *Most septic tanks must be cleaned out every few years in order to work properly.*

Labels: Inlet pipe, Ground, Cover, Scum layer, Outlet pipe, Sludge

Another Way to Clean Wastewater

If you live in an area that does not have a sewage treatment plant, your house probably uses a septic tank. **Figure 4** shows an example of a septic tank. A **septic tank** is a large underground tank that cleans the wastewater from a household. Wastewater flows from the house into the tank, where the solids sink to the bottom. Bacteria break down these wastes on the bottom of the tank. The water flows from the tank into a group of buried pipes. Then, the buried pipes, called a *drain field*, distribute the water. Distributing the water enables the water to soak into the ground.

septic tank a tank that separates solid waste from liquids and that has bacteria that break down the solid waste

Where the Water Goes

Think of some ways that you use water in your home. Do you water the lawn? Do you do the dishes? The graph in **Figure 5** shows how an average household in the United States uses water. Notice that less than 8% of the water we use in our homes is used for drinking. The rest is used for flushing toilets, doing laundry, bathing, and watering lawns and plants.

The water we use in our homes is not the only way water is used. More water is used in industry and agriculture than in homes.

✓ **Reading Check** What percentage of water in our homes is used for drinking?

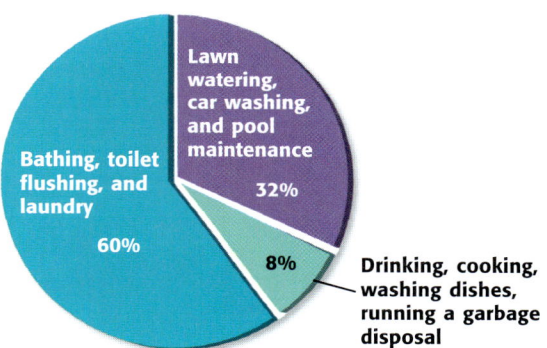

Bathing, toilet flushing, and laundry 60%

Lawn watering, car washing, and pool maintenance 32%

Drinking, cooking, washing dishes, running a garbage disposal 8%

Figure 5 *The average household in the United States uses about 100 gal of water per day. This pie graph shows some common uses of these 100 gal.*

Answer to Reading Check
Less than 8% of water in our homes is used for drinking.

Reteaching — BASIC

Saving Water at Home Have students brainstorm ways they can conserve water at home. Consolidate students' lists on the board. The list may include the following: I will be sure the dishwasher is full before it is used. If I'm washing dishes in the sink, I won't let the water run continuously. I'll turn off the water while I brush my teeth. I won't do several small loads of laundry if I can do fewer large loads instead. **LS Verbal**

Quiz — GENERAL

1. What are two sources of nonpoint-source pollution? (Sample answer: street gutters, fertilizers, eroded soils and silt from farming and logging, drainage from mines)

2. Which states get some of their groundwater from the Ogallala Aquifer? (South Dakota, Wyoming, Colorado, New Mexico, Oklahoma, Texas, Nebraska, and Kansas)

Alternative Assessment — GENERAL

Pollution Posters Provide students with poster board and markers. Direct them to create posters illustrating how nonpoint-source pollution can cause contamination of both surface water and ground water. Their diagrams should reflect the understanding that pollution of surface water can spread to ground-water. **LS Verbal** <mark>English Language Learners</mark>

MATH PRACTICE

Agriculture in Israel

From 1950 to 1980, Israel reduced the amount of water used in agriculture from 83% to 5%. Israel did so primarily by switching from overhead sprinklers to drip irrigation. A small farm uses 10,000 L of water per day for overhead sprinkler irrigation. How much water would the farm save in 1 year by using a drip irrigation system that uses 75% less water than a sprinkler system?

Group ACTIVITY — GENERAL

A Call for Conservation Divide the class into small groups. Challenge each group to write a public-service announcement to educate the public about the need to conserve water and to avoid polluting groundwater. The announcements should include definitions of point-source and nonpoint-source pollution and outline practical steps that everyone can take to protect water resources. **LS Verbal** <mark>English Language Learners</mark>

Water in Industry

About 19% of water used in the world is used for industrial purposes. Water is used to manufacture goods, cool power stations, clean industrial products, extract minerals, and generate energy for factories.

Because water resources have become expensive, many industries are trying to conserve, or use less, water. One way industries conserve water is by recycling it. In the United States, most of the water used in factories is recycled at least once. At least 90% of this recycled water can be treated and returned to surface water.

Water in Agriculture

The Ogallala aquifer is the largest known aquifer in North America. The map in **Figure 6** shows that the Ogallala aquifer runs beneath the ground through eight states, from South Dakota to Texas. The Ogallala aquifer provides water for approximately one-fifth of the cropland in the United States. Farming is the largest user of water in the Western United States. Recently, the water table in the aquifer has dropped so low that some scientists say that it would take at least 1,000 years to replenish the aquifer if it were no longer used.

Most of the water that is lost during farming is lost through evaporation and runoff. New technology, such as drip irrigation systems, has helped conserve water in agriculture. A drip irrigation system delivers small amounts of water directly to plant roots. This system allows plants to absorb the water before the water has a chance to evaporate or become runoff.

✓ Reading Check How does the drip irrigation system help conserve water?

Figure 6 *Because the Ogallala aquifer has been such a good source of groundwater, it has become overused. The water table has dropped more than 30 m in some areas.*

Answer to Reading Check

Drip irrigation systems deliver small amounts of water directly to the roots of the plant so that the plant absorbs the water before it can evaporate or run off.

Conserving Water at Home

There are many ways that people can conserve water at home. For example, many people save water by installing low-flow shower heads and low-flush toilets, because these items use much less water. To avoid watering lawns, some people plant only native plants in their yards. Native plants grow well in the local climate and don't need extra watering.

Your behavior can also help you conserve water. For example, you can take shorter showers. You can avoid running the water while brushing your teeth. And when you run the dishwasher, make sure it is full, as shown in **Figure 7.**

✔ **Reading Check** List ways in which you can conserve water in your home.

Figure 7 Run the dishwasher only when it is full.

SECTION Review

Summary

- Point-source pollution and nonpoint-source pollution are two kinds of water pollution.

- Pollutants can decrease oxygen levels and increase nitrate levels in water. These changes can cause harm to plants, animals, and humans.

- Wastewater can be treated by sewage treatment plants and septic systems.

- Water can be conserved by using only the water that is needed, by recycling water, and by using drip irrigation systems.

Using Key Terms

1. Use each of the following terms in a separate sentence: *point-source pollution, nonpoint-source pollution, sewage treatment plant,* and *septic tank.*

Understanding Key Ideas

2. Which of the following can help protect fish from acid rain?
   a. dissolved oxygen
   b. nitrates
   c. alkalinity
   d. point-source pollution

3. What type of wastewater treatment can be used for an individual home?
   a. sewage treatment plant
   b. primary treatment
   c. secondary treatment
   d. septic tank

4. Which kind of water pollution is often caused by runoff of fertilizers?

5. Describe what DO is.

6. What factors affect the level of dissolved oxygen in water?

7. Describe how water is conserved in industry.

Math Skills

8. If 25% of water used in your home is used to water the lawn and you used a total of 95 gal of water today, how many gallons of water did you use to water the lawn?

Critical Thinking

9. **Making Inferences** How do bacteria help break down the waste in water treatment plants?

10. **Applying Concepts** Other than examples listed in this section, what are some ways you can conserve water?

11. **Making Inferences** Why is it better to water your lawn at night instead of during the day?

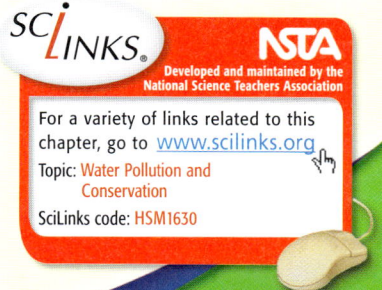

SCILINKS.

Developed and maintained by the National Science Teachers Association

For a variety of links related to this chapter, go to www.scilinks.org

Topic: Water Pollution and Conservation

SciLinks code: HSM1630

CHAPTER RESOURCES

Chapter Resource File

 • Section Quiz **GENERAL**
• Section Review **GENERAL**
• Vocabulary and Section Summary **GENERAL**
• Datasheet for Quick Lab

Technology

 Interactive Explorations CD-ROM
• Flood Bank **GENERAL**

Workbook

 Math Skills for Science
• Multiplying Whole Numbers

Water Cycle— What Goes Up . . .

Teacher's Notes

Time Required
One 45-minute class period

Lab Ratings

EASY —————————→ HARD

Teacher Prep 🔥
Student Set-Up 🔥🔥
Concept Level 🔥🔥
Clean Up 🔥

MATERIALS
The materials listed on the student page are enough for a group of 4 or 5 students.

Safety Caution
Remind students to review all safety cautions and icons before beginning this lab activity. Students should be cautioned when using a hot plate. Care should also be exercised in using the glassware. Appropriate methods should be used to dispose of broken glass.

Model-Making Lab

Water Cycle—What Goes Up . . .

Why does a bathroom mirror fog up? Where does water go when it dries up? Where does rain come from? These questions relate to the major parts of the water cycle—condensation, evaporation, and precipitation. In this activity, you will make a model of the water cycle.

OBJECTIVES

Design a model that follows the same processes as those of the water cycle.

Identify each stage of the water cycle in the model.

MATERIALS

- beaker
- gloves, heat-resistant
- graduated cylinder
- hot plate
- plate, glass, or watch glass
- tap water, 50 mL
- tongs or forceps

SAFETY

Procedure

1. Use the graduated cylinder to pour 50 mL of water into the beaker. Note the water level in the beaker.

2. Put on your safety goggles and gloves. Place the beaker securely on the hot plate. Turn the heat to medium, and bring the water to a boil.

3. While waiting for the water to boil, practice picking up and handling the glass plate or watch glass with the tongs. Hold the glass plate a few centimeters above the beaker, and tilt it so that the lowest edge of the glass is still above the beaker.

4. Observe the glass plate as the water in the beaker boils. Record the changes you see in the beaker, in the air above the beaker, and on the glass plate held over the beaker. Write down any changes you see in the water.

CHAPTER RESOURCES

Chapter Resource File

- Datasheet for Chapter Lab
- Lab Notes and Answers

Technology

Classroom Videos
- Lab Video

LabBook
- Clean Up Your Act

Analyze the Results

1. Sketches should resemble the illustration of the water cycle at the bottom of this page. Students should include labels for condensation, evaporation, and precipitation.

2. Sample answer: The water level is lower at the end of the experiment because some of the water evaporated.

Draw Conclusions

3. Sample answer: The mass would have changed slightly. Because some of the steam escaped, the mass of the water in the beaker would decrease.

4. Sample answer: The model is similar to the Earth's water cycle because evaporation, condensation, and precipitation occurred. Accept all reasonable depictions of the water cycle.

5. Sample answer: Much of the energy in the water cycle is stored in bodies of water, such as the oceans. Thermal energy is also stored in the atmosphere and the Earth's land surface.

Applying Your Data

No, minerals and salts are not part of the water cycle. As water evaporates, minerals and salts are left behind as deposits.

5. Continue until you have observed steam rising off the water, the glass plate becoming foggy, and water dripping from the glass plate.

6. Carefully set the glass plate on a counter or other safe surface as directed by your teacher.

7. Turn off the hot plate, and allow the beaker to cool. Move the hot beaker with gloves or tongs if you are directed to do so by your teacher.

Analyze the Results

1. **Constructing Charts** Copy the illustration shown above. On your sketch, draw and label the water cycle as it happened in your model. Include arrows and labels for *evaporation, condensation,* and *precipitation.*

2. **Analyzing Results** Compare the water level in the beaker now with the water level at the beginning of the experiment. Was there a change? Explain why or why not.

Draw Conclusions

3. **Making Predictions** If you had used a scale or a balance to measure the mass of the water in the beaker before and after this activity, would the mass have changed? Explain.

4. **Analyzing Charts** How is your model similar to the Earth's water cycle? On your sketch of the illustration, label where the processes shown in the model reflect the Earth's water cycle.

5. **Drawing Conclusions** When you finished this experiment, the water in the beaker was still hot. What stores much of the energy in the Earth's water cycle?

Applying Your Data

As rainwater runs over the land, the water picks up minerals and salts. Do these minerals and salts evaporate, condense, and precipitate as part of the water cycle? Where do they go?

Students' sketches should resemble the water cycle illustrated here:

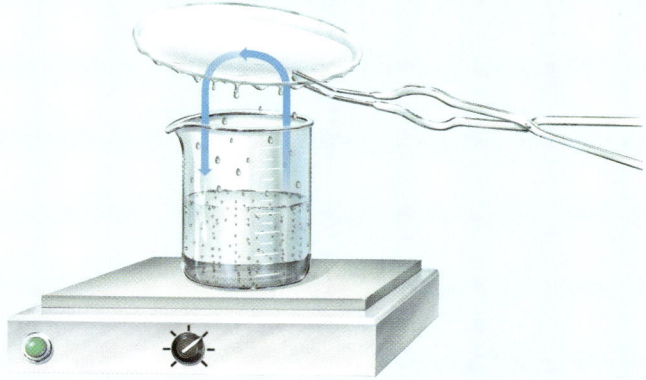

Assignment Guide

SECTION	QUESTIONS
1	1, 4, 7–8, 18, 21–24
2	2, 9
3	3, 5, 10, 13–16
4	6, 11–12, 19–20
1 and 2	17

ANSWERS

Using Key Terms

1. A stream that flows into a lake or into a larger stream is a tributary.

2. The area along a river that forms from sediment deposited when the river overflows is a floodplain.

3. A rock's ability to let water through it is called permeability.

4. Sample answer: A divide is the boundary between two drainage areas. A watershed is a drainage area.

5. Sample answer: An artesian spring is a spring that flows through a natural crack in the cap rock. A well is a human made hole that reaches below the water table.

6. Sample answer: Point-source pollution is pollution that enters the water from a known single source, such as a factory. Nonpoint-source pollution is pollution that comes from many sources, such as a combination of fertilizer and pesticide in groundwater.

USING KEY TERMS

The statements below are false. For each statement, replace the underlined term to make a true statement.

1 A stream that flows into a lake or into a larger stream is a <u>water cycle</u>.

2 The area along a river that forms from sediment deposited when the river overflows is a <u>delta</u>.

3 A rock's ability to let water pass through it is called <u>porosity</u>.

For each pair of terms, explain how the meanings of the terms differ.

4 *divide* and *watershed*

5 *artesian springs* and *wells*

6 *point-source pollution* and *nonpoint-source pollution*

UNDERSTANDING KEY IDEAS

Multiple Choice

7 Which of the following processes is not part of the water cycle?
   a. evaporation
   b. percolation
   c. condensation
   d. deposition

8 Which features are common in youthful river channels?
   a. meanders
   b. flood plains
   c. rapids
   d. sandbars

9 Which depositional feature is found at the coast?
   a. delta
   b. flood plain
   c. alluvial fan
   d. placer deposit

10 Caves are mainly a product of
   a. erosion by rivers.
   b. river deposition.
   c. water pollution.
   d. erosion by groundwater.

11 Which of the following is necessary for aquatic life to survive?
   a. dissolved oxygen
   b. nitrates
   c. alkalinity
   d. point-source pollution

12 During primary treatment at a sewage treatment plant,
   a. water is sent to an aeration tank.
   b. water is mixed with bacteria and oxygen.
   c. dirty water is passed through a large screen.
   d. water is sent to a settling tank where chlorine is added.

Short Answer

13 Identify and describe the location of the water table.

14 Explain how surface water enters an aquifer.

15 Why are caves usually found in limestone-rich regions?

Understanding Key Ideas

7. d
8. c
9. a
10. d
11. a
12. c
13. A water table is the upper surface of underground water and the upper boundary of the zone of saturation.

14. Aquifers are replenished in the recharge zone. The recharge zone is an area where a permeable rock layer allows water to percolate into the aquifer.

15. Limestone is made of calcium carbonate, which dissolves easily in water. Groundwater dissolves the limestone and produces caves.

CRITICAL THINKING

16. Concept Mapping Use the following terms to create a concept map: *zone of aeration, zone of saturation, water table, gravity, porosity,* and *permeability.*

17. Identifying Relationships What is water's role in erosion and deposition?

18. Analyzing Processes What are the features of a river channel that has a steep gradient?

19. Analyzing Processes Why is groundwater hard to clean?

20. Evaluating Conclusions How can water be considered both a renewable and a nonrenewable resource? Give an example of each case.

21. Analyzing Processes Does water vapor lose or gain energy during the process of condensation? Explain.

INTERPRETING GRAPHICS

The hydrograph below illustrates data collected on river flow during field investigations over a period of 1 year. The discharge readings are from the Yakima River, in Washington. Use the hydrograph below to answer the questions that follow.

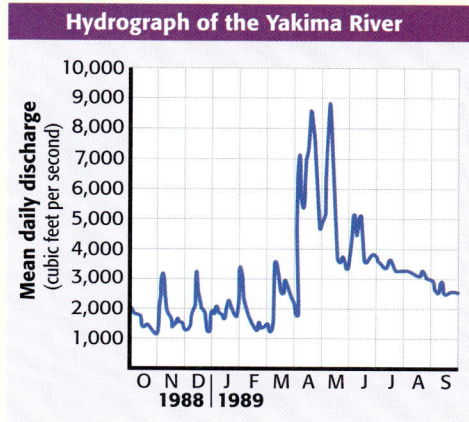

Hydrograph of the Yakima River

22. In which months is there the highest river discharge?

23. Why is there such a high river discharge during these months?

24. What might cause the peaks in river discharge between November and March?

16. 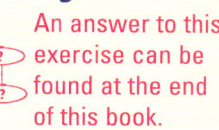 An answer to this exercise can be found at the end of this book.

17. Water flows across a landscape, eroding, transporting, and depositing material. Water is an agent of erosion and deposition.

18. A river channel that has a steep gradient is straight and narrow with rapids, waterfalls, and V-shaped valleys.

19. Once groundwater becomes polluted, it is hard to clean because it is not at the surface. Also, it moves very slowly and will therefore take a long time to clean.

20. Water is a renewable resource when it can be replaced or recycled. Rain is a renewable resource. Water is nonrenewable when it is consumed faster than it can be replenished. An example is the Ogallala aquifer.

21. Answers may vary. Students should conclude that water loses energy during the process of condensation. As water cools and condenses it loses heat energy.

Interpreting Graphics

22. April and May

23. Accept all reasonable responses. Sample answer: spring snowmelt from the mountains and high rainfall

24. Accept all reasonable responses. Sample answer: winter storms or thaws

CHAPTER RESOURCES

Chapter Resource File

- Chapter Review GENERAL
- Chapter Test A GENERAL
- Chapter Test B ADVANCED
- Chapter Test C SPECIAL NEEDS
- Vocabulary Activity GENERAL

Workbooks

Study Guide
- Assessment resources are also available in Spanish.

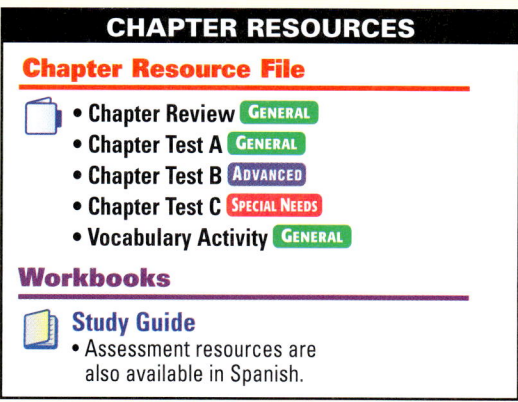

Standardized Test Preparation

Teacher's Note

To provide practice under more realistic testing conditions, give students 20 minutes to answer all of the questions in this Standardized Test Preparation.

MISCONCEPTION ALERT

Answers to the standardized test preparation can help you identify student misconceptions and misunderstandings.

READING

Passage 1

1. B
2. F
3. B

TEST DOCTOR

Question 3: Students may choose answer A because the text mentions that Old Faithful erupts every 60 to 70 minutes. However, the time in the text is approximate, and the statement in answer A indicates that the eruption takes place every 60 minutes exactly, which is not true.

READING

Read each of the passages below. Then, answer the questions that follow each passage.

Passage 1 In parts of Yellowstone National Park, boiling water from deep in the ground blasts into the sky. These blasts of steam come from lakes of strange-colored boiling mud that gurgle and hiss. These features are called geysers. Yellowstone's most popular geyser is named Old Faithful. It is given this name because it erupts every 60 min to 70 min without fail. A geyser is formed when a narrow vent connects one or more underground chambers to Earth's surface. These underground chambers are heated by nearby molten rock. As underground water flows into the vent and chambers, it is heated above 100°C. This superheated water quickly turns to steam and explodes, projecting <u>scalding</u> water 60 m into the air. And Old Faithful erupts right on schedule!

1. In the passage, what does *scalding* mean?
   - **A** muddy
   - **B** burning
   - **C** gurgling
   - **D** steaming

2. According to the passage, what happens to underground water when geysers form?
   - **F** It is heated by molten rock.
   - **G** It is cycled to Earth's center.
   - **H** It travels 60 m through vents.
   - **I** It is poured into volcanoes.

3. Which of the following is a fact in the passage?
   - **A** Old Faithful erupts every 60 min.
   - **B** Old Faithful is located in Yellowstone National Park.
   - **C** There are six geysers at Yellowstone National Park.
   - **D** Molten rock explodes from geysers.

Passage 2 In the Mississippi Delta, long-legged birds step lightly through the marsh and hunt fish or frogs for breakfast. Hundreds of species of plants and animals start another day in this fragile ecosystem. This delta ecosystem is in danger of being destroyed. The threat comes from efforts to make the river more useful. Large portions of the river bottom were <u>dredged</u> to deepen the river for ship traffic. Underwater channels were built to control flooding. What no one realized was that sediments that once formed new land now passed through the channels and flowed out into the ocean. Those river sediments had once replaced the land that was lost every year to erosion. Without them, the river can't replace land lost to erosion. So, the Mississippi River Delta is shrinking. By 1995, more than half of the wetlands were already gone—swept out to sea by waves along the Louisiana coast.

1. In the passage, what does *dredged* mean?
   - **A** moved to the side
   - **B** circulated
   - **C** cleaned
   - **D** scooped up

2. Based on the passage, which of the following statements about the Mississippi River is true?
   - **F** The river never floods.
   - **G** The river is not wide enough for ships.
   - **H** The river's delicate ecosystem is in danger.
   - **I** The river is disappearing.

3. Which of the following is a fact in the passage?
   - **A** By 1995, more than half of the Mississippi River was gone.
   - **B** Underwater channels controlled flooding.
   - **C** Channels help form new land.
   - **D** Sediment cannot replace lost land.

Passage 2

1. D
2. H
3. B

Question 3: Students who do not thoroughly read the passage may choose answer A. They may incorrectly think that half the river system and not the wetlands has been lost.

The chart below shows four wells drilled at different depths. Use the chart below to answer the questions that follow.

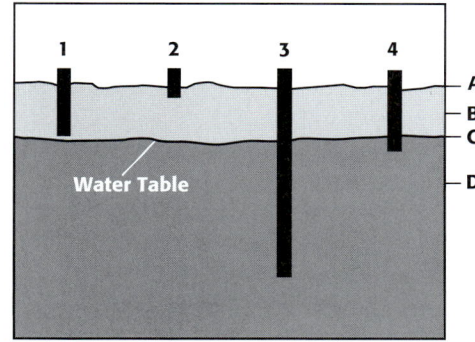

Water Table

1. A well-drilling company offers the four types of wells shown in the chart. Which well is most likely to be a reliable source of groundwater?

A 1
B 2
C 3
D 4

2. If the area experienced heavy rains, toward which level would the water table move?

F The water table would move toward level B.
G The water table would move toward level D.
H The water table would stay at level C.
I The water table will be gone.

3. If the water table moves to level D, which wells will still be able to provide water?

A all wells
B wells 1 and 2
C well 3
D wells 3 and 4

4. Which well is most likely to be an unreliable source of groundwater?

F 1
G 2
H 3
I 4

Read each question below, and choose the best answer.

1. A river flows at a speed of 10 km/h. If a boat travels upstream at a speed of 15 km/h, how far will it travel in 3 h?

A 10 km
B 15 km
C 20 km
D 25 km

2. Water contamination is often measured in parts per million (ppm). If the concentration of a pollutant is 5 ppm, there are 5 parts of the pollutant in 1 million parts of water. If the concentration of gasoline is 3 ppm in 2,000,000 L of water, how many liters of gasoline are in the water?

F 3 L
G 6 L
H 9 L
I 10 L

3. One family uses 70 L of water a day for showering. If everyone in the family agreed to shorten his or her shower from 10 min to 5 min, how many liters of water would be saved each day?

A 5 L
B 10 L
C 35 L
D 70 L

4. A family uses 800 L of water per day. Of those 800 L, 200 L are used for flushing the toilet. Calculate the percentage of water that the family uses to flush the toilet.

F 25%
G 30%
H 50%
I 60%

5. A river flows at a speed of 8 km/h. If you floated on a raft in this river, how far will you have traveled after 5 h?

A 5 km
B 16 km
C 40 km
D 80 km

Standardized Test Preparation

1. C
2. F
3. C
4. G

TEST DOCTOR

Question 2: Students should understand that the water table is the upper surface of the groundwater. When it rains, the ground absorbs more water, and the water level rises, so the water table goes up.

1. B
2. G
3. C
4. F
5. C

TEST DOCTOR

Question 2: When measuring in parts per million, the units of the solute is the same as the units of the solvent. In this case, the solvent is water, and pollutants are present at 3 ppm in 2,000,000 L of water. So, there are 6 L of pollutant in 2,000,000 L of water.

Science in Action

Weird Science

Discussion ——— GENERAL

Scientists can learn alot by studying samples of water from Lake Vostok. However, investigating the lake will come with a price: damage to a completely untouched environment. It is currently impossible to avoid contaminating Lake Vostok in the process of gathering samples. Have students debate whether the information gained by studying Lake Vostok is worth the risk of contaminating the lake. **LS Verbal**

Scientific Discoveries

Background

Ankarana is a limestone formation known as karst topography. Karst topography describes a region where the effects of chemical weathering due to groundwater are clearly visible at the surface. Groundwater has carved long, mostly unexplored caves into the rock. There are human bones that date back as far as 750 CE in the caves. Some of these bones belong to Antakarana kings. The Antakarana people consider the caves sacred. They will not enter the caves except to conduct annual ceremonies in honor of the kings.

Science in Action

Weird Science

Secret Lake

Would you believe there is a freshwater lake more than 3 km below an Antarctic glacier near the South Pole? It is surprising that Lake Vostok can remain in a liquid state at a place where the temperature can fall below −50°C. Scientists believe that the intense pressure from the overlying ice heats the lake and keeps it from freezing. Geothermal energy, which is the energy within the surface of the Earth, also contributes to warmer temperatures. The other unique thing about Lake Vostok is the discovery of living microbes under the glacier that covers the lake!

Language Arts ACTIVITY

Look up the word *geothermal* in the dictionary. What is the meaning of the roots *geo-* and *-thermal*? Find other words in the dictionary that begin with the root *geo-*.

Vostok Station

Drilled core (3,623 m down)   Glacial ice

Lake ice

Lake Vostok (at least 500 m deep)

Sediment

Scientific Discoveries

Sunken Forests

Imagine having your own little secret forest. In Ankarana National Park, in Madagascar, there are plenty of them. Within the limestone mountain of the park, caves have formed from the twisting path of the flowing groundwater. In many places in the caves, the roof has collapsed to form a sinkhole. The light that now shines through the collapsed roof of the cave has allowed miniature sunken forests to grow. Each sunken forest has unique characteristics. Some have crocodiles. Others have blind cavefish. You can even find some species that can't be found anywhere else in the world!

Social Studies ACTIVITY

Find out how Madagascar's geography contributes to the biodiversity of the island nation. Make a map of the island that highlights some of the unique forms of life found there.

Answer to Language Arts Activity

The root *geo-* means "earth," and *-thermal* means "heat." Other words that begin with the prefix *geo-* include *geology* and *geometry*. *Geology* is the study of the Earth. *Geometry* literally means "to measure the Earth."

Answer to Social Studies Activity

Answers may vary. Accept any reasonable depiction of Madagascar's biodiversity. Answers should include many species of lemur and a wide variety of flora.

People in Science

Rita Colwell

A Water Filter for All Did you ever drink a glass of water through a piece of cloth? Dr. Rita Colwell, director of the National Science Foundation, has found that filtering drinking water through a cloth can actually decrease the number of disease-causing bacteria in the water. This discovery is very important for the people of Bangladesh, where deadly outbreaks of cholera are frequent. People are usually infected by the cholera bacteria by drinking contaminated water. Colwell knew that filtering the water would remove the bacteria. The water would then be safe to drink. Unfortunately, filters were too expensive for most of the people to buy. Colwell tried filtering the water with a sari. A sari is a long piece of colorful cloth that many women in Bangladesh wear as skirtlike cloth. Filtering the water with the sari cloth did the trick. The amount of cholera bacteria in the water was reduced. Fewer people contracted cholera, and many lives were saved!

Math  Activity

With the cloth water-filter method, there was a 48% reduction in the occurrence of cholera. If there were 125 people out of 100,000 who contracted cholera before the cloth-filter method was used, how many people per 100,000 contracted cholera after using the cloth-filter method?

To learn more about these Science in Action topics, visit **go.hrw.com** and type in the keyword **HZ5DEPF**.

Current Science

Check out Current Science® articles related to this chapter by visiting **go.hrw.com**. Just type in the keyword **HZ5CS11**.

Exploring the Oceans
Chapter Planning Guide

Compression guide:
To shorten instruction because of time limitations, omit the Chapter Lab.

OBJECTIVES	LABS, DEMONSTRATIONS, AND ACTIVITIES	TECHNOLOGY RESOURCES
PACING • 90 min pp. 36–45 **Chapter Opener**	**SE** Start-up Activity, p. 37 `GENERAL`	**OSP** Parent Letter ■ `GENERAL` **CD** Student Edition on CD-ROM **CD** Guided Reading Audio CD ■ **TR** Chapter Starter Transparency* **VID** Brain Food Video Quiz
Section 1 Earth's Oceans • List the major divisions of the global ocean. • Describe the history of Earth's oceans. • Identify the properties of ocean water. • Describe the interactions between the ocean and the atmosphere.	**TE** Activity Ocean Size, p. 38 `GENERAL` **TE** Activity Diagramming Temperature Zones, p. 41 `BASIC` **SE** Connection to Geology Submarine Volcanoes, p. 42 ◆ `GENERAL` **TE** Group Activity Making Models, p. 42 ◆ `BASIC` **TE** Connection Activity Geography, p. 42 `GENERAL` **TE** Activity Modeling the Water Cycle, p. 43 ◆ `BASIC`	**CRF** Lesson Plans* **TR** Bellringer Transparency* **TR** Divisions of the Global Ocean* **TR** Ocean Salinity* **TR** The Ocean and the Water Cycle* **SE** Internet Activity, p. 44 `GENERAL` **CD** Science Tutor
PACING • 45 min pp. 46–51 **Section 2 The Ocean Floor** • Describe technologies for studying the ocean floor. • Identify the two major regions of the ocean floor. • Classify subdivisions and features of the two major regions of the ocean floor.	**TE** Connection Activity Language Arts, p. 47 ◆ `ADVANCED` **TE** Connection Activity Art, p. 49 `GENERAL` **SE** Connection to Social Studies The JASON Project, p. 50 `GENERAL` **SE** Model-Making Lab Probing the Depths, p. 70 ◆ `GENERAL` **LB** Calculator-Based Labs Ocean Floor Mapping* ◆ `ADVANCED`	**CRF** Lesson Plans* **TR** Bellringer Transparency* **TR** How Sonar Works* **TR** Revealing the Ocean Floor: A* **TR** Revealing the Ocean Floor: B* **VID** Lab Videos for Earth Science **CD** Science Tutor
PACING • 45 min pp. 52–57 **Section 3 Life in the Ocean** • Identify the three groups of marine life. • Describe the two main ocean environments. • Identify the ecological zones of the benthic and pelagic environments.	**TE** Group Activity Classifying, p. 52 `GENERAL` **TE** Group Activity Ocean Zones and Organisms, p. 54 ◆ `GENERAL` **SE** Connection to Language Arts Water, Water, Everywhere, p. 56 `GENERAL` **LB** Whiz-Bang Demonstrations Foul Play* ◆ `GENERAL` **LB** EcoLabs & Field Activities Operation Oil-Spill Cleanup* ◆ `GENERAL` **LB** Long-Term Projects & Research Ideas Your Very Own Underwater Theme Park* `ADVANCED`	**CRF** Lesson Plans* **TR** Bellringer Transparency* **TR** The Three Groups of Marine Life* **TR** LINK TO LIFE SCIENCE Four Parts of Natural Selection* **CRF** SciLinks Activity* `GENERAL` **CD** Interactive Explorations CD-ROM Sea Sick `GENERAL`
PACING • 45 min pp. 58–63 **Section 4 Resources from the Ocean** • List two ways of harvesting the ocean's living resources. • Identify three nonliving resources in the ocean. • Describe the ocean's energy resources.	**TE** Group Activity Brainstorming, p. 58 `GENERAL` **TE** Connection Activity Real World, p. 59 `ADVANCED` **TE** Connection Activity Real World, p. 60 `GENERAL` **TE** Group Activity Public Service Announcement, p. 60 `GENERAL` **SE** Quick Lab Desalination Plant, p. 61 ◆ `GENERAL` **LB** Inquiry Labs Surf's Up!* ◆ `GENERAL`	**CRF** Lesson Plans* **TR** Bellringer Transparency* **CD** Science Tutor
PACING • 45 min pp. 64–69 **Section 5 Ocean Pollution** • Explain the difference between point-source pollution and nonpoint-source pollution. • Identify three different types of point-source ocean pollution. • Describe what is being done to control ocean pollution.	**TE** Activity Cleaning Up an Oil Spill, p. 64 `GENERAL` **TE** Activity Nonpoint-source Pollution, p. 65 `BASIC` **TE** Connection Activity Math, p. 66 `GENERAL` **TE** Activity Ocean Pollution Awareness, p. 67 `GENERAL` **TE** Group Activity *Exxon Valdez*, p. 67 `GENERAL` **SE** School-to-Home Activity Coastal Cleanup, p. 68 `GENERAL` **TE** Group Activity Coastal Campaign, p. 68 `GENERAL` **SE** Model-Making Lab Investigating an Oil Spill, p. 112 ◆ `GENERAL`	**CRF** Lesson Plans* **TR** Bellringer Transparency* **CD** Science Tutor

PACING • 90 min

CHAPTER REVIEW, ASSESSMENT, AND STANDARDIZED TEST PREPARATION

CRF Vocabulary Activity* `GENERAL`
SE Chapter Review, pp. 72–73 `GENERAL`
CRF Chapter Review* ■ `GENERAL`
CRF Chapter Tests A* ■ `GENERAL`, B* `ADVANCED`, C* `SPECIAL NEEDS`
SE Standardized Test Preparation, pp. 74–75 `GENERAL`
CRF Standardized Test Preparation* `GENERAL`
CRF Performance-Based Assessment* `GENERAL`
OSP Test Generator `GENERAL`
CRF Test Item Listing* `GENERAL`

Online and Technology Resources

Visit **go.hrw.com** for a variety of free resources related to this textbook. Enter the keyword **HA5OCE.**

Students can access interactive problem-solving help and active visual concept development with the *Holt Science and Technology* Online Edition available at **www.hrw.com.**

Guided Reading Audio CD
Also in Spanish
A direct reading of each chapter for auditory learners, reluctant readers, and Spanish-speaking students.

Science Tutor CD-ROM
Excellent for remediation and test practice.

SKILLS DEVELOPMENT RESOURCES	SECTION REVIEW AND ASSESSMENT	STANDARDS CORRELATIONS
SE Pre-Reading Activity, p. 36 `GENERAL` **OSP** Science Puzzlers, Twisters & Teasers* `GENERAL`		National Science Education Standards UCP 2, 5; SAI 1; ST 2; SPSP 5
CRF Directed Reading A* ■ `BASIC`, B* `SPECIAL NEEDS` **CRF** Vocabulary and Section Summary* ■ `GENERAL` **SE** Reading Strategy Discussion, p. 38 `GENERAL` **TE** Reading Strategy Mnemonics, p. 39 `GENERAL` **TE** Inclusion Strategies, p. 41 ◆	**SE** Reading Checks, pp. 39, 40, 42, 44 `GENERAL` **TE** Homework, p. 43 `ADVANCED` **TE** Reteaching, p. 44 `BASIC` **TE** Quiz, p. 44 `GENERAL` **TE** Alternative Assessment, p. 44 `ADVANCED` **SE** Section Review,* p. 45 `GENERAL` **CRF** Section Quiz* ■ `GENERAL`	UCP 1, 2, 3; ES 1b, 1f, 1g, 1h, 1j, 2a
CRF Directed Reading A* ■ `BASIC`, B* `SPECIAL NEEDS` **CRF** Vocabulary and Section Summary* ■ `GENERAL` **SE** Reading Strategy Reading Organizer, p. 46 `GENERAL` **MS** Math Skills for Science Multiplying Whole Numbers* `GENERAL` **MS** Math Skills for Science Multiplying and Dividing Fractions* `GENERAL`	**SE** Reading Checks, pp. 47, 48, 49, 50 `GENERAL` **TE** Homework, p. 48 `ADVANCED` **TE** Reteaching, p. 50 `BASIC` **TE** Quiz, p. 50 `GENERAL` **TE** Alternative Assessment, p. 50 `GENERAL` **SE** Section Review,* p. 51 `GENERAL` **CRF** Section Quiz* ■ `GENERAL`	UCP 2, 3; SAI 1, 2; ST 2; SPSP 5; HNS 1, 3; ES 1b, 1c; *Chapter Lab:* UCP 2, 3; SAI 1, 2; ST 2; SPSP 5; HNS 1
CRF Directed Reading A* ■ `BASIC`, B* `SPECIAL NEEDS` **CRF** Vocabulary and Section Summary* ■ `GENERAL` **SE** Reading Strategy Mnemonics, p. 52 `GENERAL` **TE** Inclusion Strategy, p. 55 **CRF** Reinforcement Worksheet The Ocean's Environment* `BASIC`	**SE** Reading Checks, pp. 53, 55, 56 `GENERAL` **TE** Homework, p. 53, 54 `GENERAL` **TE** Reteaching, p. 56 `BASIC` **TE** Quiz, p. 56 `GENERAL` **TE** Alternative Assessment, p. 56 `GENERAL` **SE** Section Review,* p. 57 `GENERAL` **CRF** Section Quiz* ■ `GENERAL`	UCP 1
CRF Directed Reading A* ■ `BASIC`, B* `SPECIAL NEEDS` **CRF** Vocabulary and Section Summary* ■ `GENERAL` **SE** Reading Strategy Paired Summarizing, p. 58 `GENERAL` **TE** Reading Strategy Prediction Guide, p. 59 `BASIC` **CRF** Reinforcement Worksheet The Oceans and Us* `BASIC` **CRF** Critical Thinking Chain Reaction* `ADVANCED`	**SE** Reading Checks, pp. 59, 60, 61, 63 `GENERAL` **TE** Homework, p. 60 `GENERAL` **TE** Reteaching, p. 62 `BASIC` **TE** Quiz, p. 62 `GENERAL` **TE** Alternative Assessment, p. 62 `GENERAL` **SE** Section Review,* p. 63 `GENERAL` **CRF** Section Quiz* ■ `GENERAL`	SAI 1; ST 2; SPSP 2, 4, 5; HNS 1
CRF Directed Reading A* ■ `BASIC`, B* `SPECIAL NEEDS` **CRF** Vocabulary and Section Summary* ■ `GENERAL` **SE** Reading Strategy Reading Organizer, p. 64 `GENERAL`	**SE** Reading Checks, pp. 65, 67, 69 `GENERAL` **TE** Homework, p. 65 `GENERAL` **TE** Reteaching, p. 68 `BASIC` **TE** Quiz, p. 68 `GENERAL` **TE** Alternative Assessment, p. 68 `GENERAL` **SE** Section Review,* p. 69 `GENERAL` **CRF** Section Quiz* ■ `GENERAL`	ST 2; SPSP 2, 4, 5; *LabBook:* UPC 2, 3; SAI 1; SPSP 2, 3, 4; HNS 1

One-Stop Planner® CD-ROM

This convenient CD-ROM includes:
- Lab Materials QuickList Software
- Holt Calendar Planner
- Customizable Lesson Plans
- Printable Worksheets
- ExamView® Test Generator

cnnstudentnews.com

Find the latest news, lesson plans, and activities related to important scientific events.

www.scilinks.org

Maintained by the **National Science Teachers Association.** See Chapter Enrichment pages for a complete list of topics.

Check out *Current Science* articles and activities by visiting the HRW Web site at **go.hrw.com.** Just type in the keyword **HZ5CS13T.**

Classroom Videos

- **Lab Videos** demonstrate the chapter lab.
- **Brain Food Video Quizzes** help students review the chapter material.
- **CNN Videos** bring science into your students' daily life.

Chapter Resources

Visual Resources

CHAPTER STARTER TRANSPARENCY

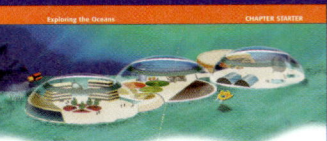

BELLRINGER TRANSPARENCIES

TEACHING TRANSPARENCIES

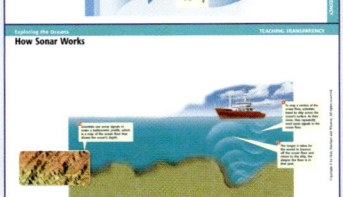

TEACHING TRANSPARENCIES

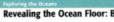
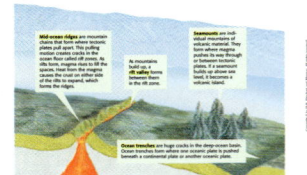
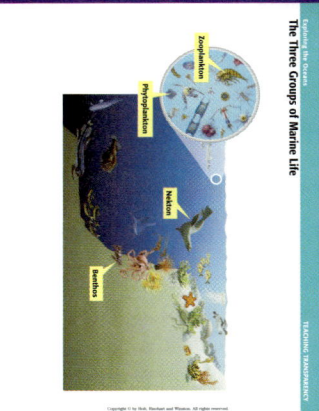

CONCEPT MAPPING TRANSPARENCY

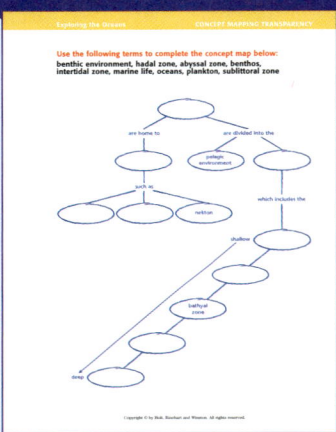

Planning Resources

LESSON PLANS

PARENT LETTER

ALSO IN SPANISH

TEST ITEM LISTING

One-Stop Planner® CD-ROM

This CD-ROM includes all of the resources shown here and the following time-saving tools:

- *Lab Materials QuickList Software*
- *Customizable lesson plans*
- *Holt Calendar Planner*
- *The powerful ExamView® Test Generator*

For a preview of available worksheets covering math and science skills, see pages T12–T19. All of these resources are also on the One-Stop Planner®.

Meeting Individual Needs

DIRECTED READING A
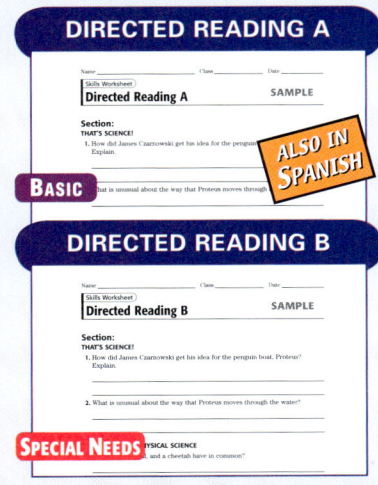
BASIC — ALSO IN SPANISH

DIRECTED READING B
SPECIAL NEEDS

VOCABULARY ACTIVITY
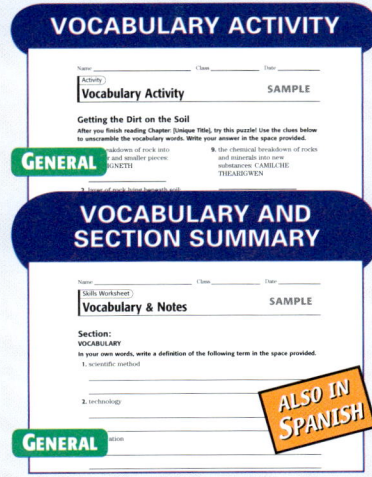
GENERAL

VOCABULARY AND SECTION SUMMARY
GENERAL — ALSO IN SPANISH

REINFORCEMENT
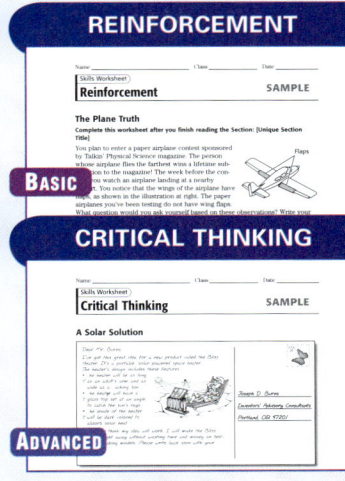
BASIC

CRITICAL THINKING
ADVANCED

SCILINKS ACTIVITY

GENERAL

SCIENCE PUZZLERS, TWISTERS & TEASERS
GENERAL

Labs and Activities

ECOLABS & FIELD ACTIVITIES
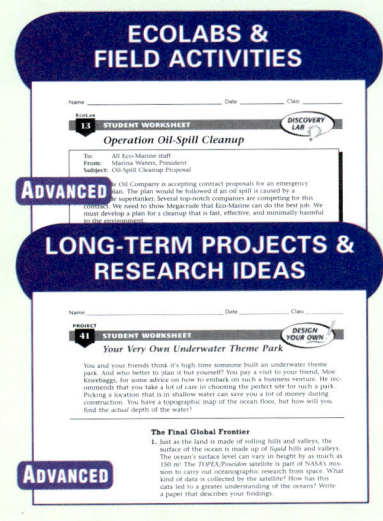
ADVANCED

LONG-TERM PROJECTS & RESEARCH IDEAS
ADVANCED

WHIZ-BANG DEMONSTRATIONS
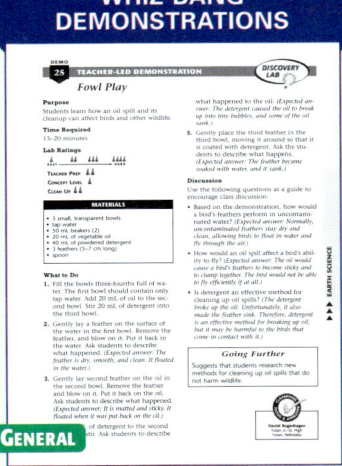
GENERAL

INQUIRY LABS
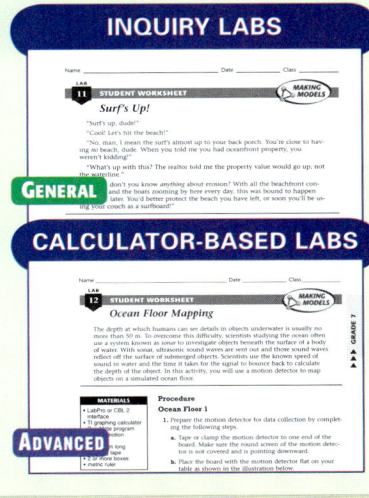
GENERAL

CALCULATOR-BASED LABS
ADVANCED

DATASHEETS FOR QUICKLABS
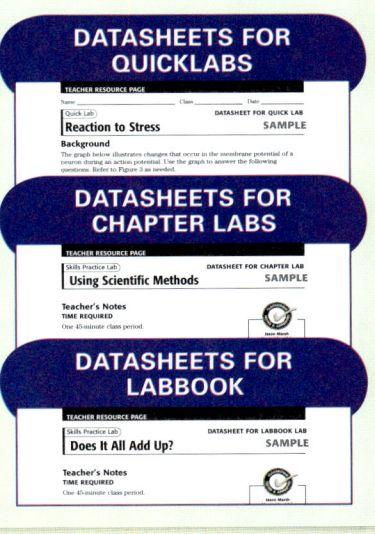

DATASHEETS FOR CHAPTER LABS

DATASHEETS FOR LABBOOK

Review and Assessments

SECTION QUIZ
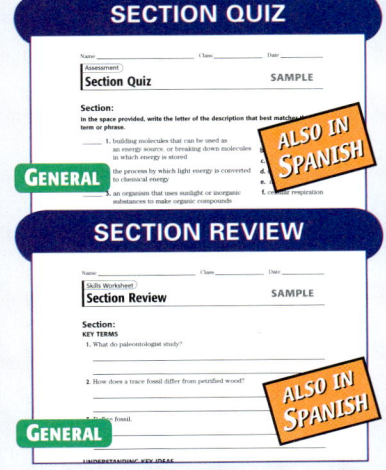
GENERAL — ALSO IN SPANISH

SECTION REVIEW
GENERAL — ALSO IN SPANISH

CHAPTER REVIEW
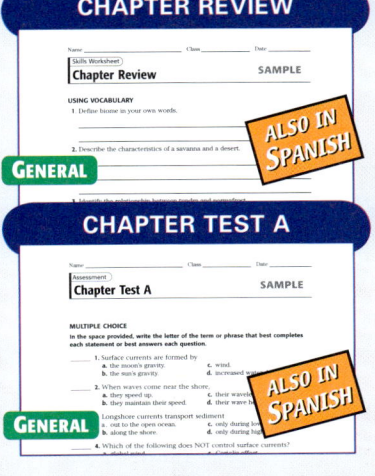
GENERAL — ALSO IN SPANISH

CHAPTER TEST A
GENERAL — ALSO IN SPANISH

CHAPTER TEST B
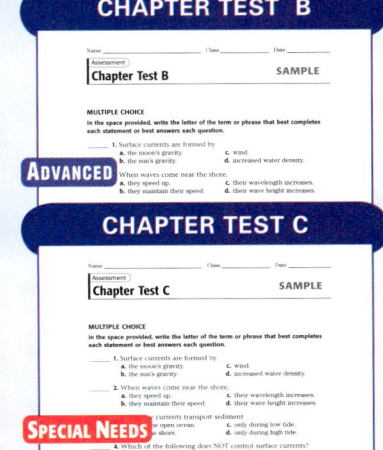
ADVANCED

CHAPTER TEST C
SPECIAL NEEDS

STANDARDIZED TEST PREPARATION
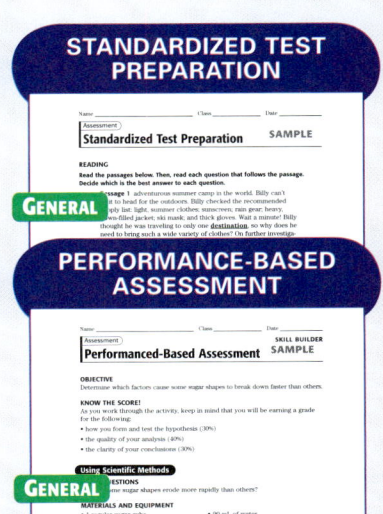
GENERAL

PERFORMANCE-BASED ASSESSMENT
GENERAL

This Chapter Enrichment provides relevant and interesting information to expand and enhance your presentation of the chapter material.

Section 1

Earth's Oceans
The Global Ocean

- Historically, the global ocean was divided into five oceans: the Atlantic, Pacific, Indian, Arctic, and Antarctic Oceans. Today, most oceanographers agree that the Antarctic Ocean is actually the southernmost section of the Atlantic, Pacific, and Indian Oceans.

The Mariana Trench

- In 1960, the *Trieste,* a small submersible designed to explore the ocean to great depths, set out on a voyage that until then had only been imagined: it descended into the Mariana Trench, the deepest known place on Earth. As Jacques Piccard and a companion descended in the *Treiste,* they were surprised to feel abrupt changes between the ocean's temperature layers. Every time the vessel reached the boundary between two layers, the *Trieste* seemed to stop as though it had reached the ocean floor.

Is That a Fact!

- ◆ The average depth of each ocean is as follows: Arctic: 1,038 m; Atlantic: 3,735 m; Indian: 3,872 m; and Pacific: 4,188 m
- ◆ The average depth for all the oceans is about 3,800 m.

Section 2

The Ocean Floor
The Renewal of a Planet

- Submersible missions such as those of *Alvin* have enabled oceanographers to witness the creation of oceanic crust and the forces that drive tectonic plate movement. By observing molten rock welling up into the ocean, they have seen how new sea floor forms and have gained a better understanding of how the continents drift apart. Trained to "read" rock formations, these scientists use their observations to reconstruct Earth's history.

Is That a Fact!

- ◆ The oceans' deep-water sound channels carry sound waves for hundreds of kilometers. Whales and other marine animals take advantage of these properties to communicate over long ranges and to search for food. Whales communicate with clicks, whistles, squeaks, and songs that convey information. Scientists aren't sure what these songs mean but do know that whales can communicate at distances as great as 1,600 km!

Section 3

Life in the Ocean
A World of Its Own

- In 1977, scientists aboard *Alvin* witnessed an astonishing new world around deep-sea vents. Exploring hydrothermal vents 320 km off the Galápagos Islands, these explorers saw an amazing multitude of unusual marine populations—giant clams and worms, fish, and crabs— gathered in an abyssal oasis.

- Using a special claw attachment, scientists harvested samples of the marine life they found. When they analyzed their specimens, they were in for a surprise—the water from around the specimans smelled like rotten eggs! This smell came from hydrogen sulfide dissolved in the water around the vents.

- The scientists discovered that certain marine bacteria thrive on the hydrogen sulfide released by vents in the ocean floor. These bacteria are food for the other marine creatures, and they are part of a food chain that does not rely on photosynthesis for energy.

Section 4

Resources from the Ocean

Food for Thought

- Fish are an invaluable ocean resource, providing a significant percentage of the world's protein needs every year. About 75 million tons of fish are harvested from the ocean each year. But many fish populations have been depleted by overfishing.

- Concerned that too few fish will remain to breed, scientists determine the maximum sustainable yield, or the amount of fish that can be harvested each year without jeopardizing future catches. Using the scientists' guidelines, governments sometimes impose fishing restrictions to manage fish populations. Many people are working to ensure that the world can continue to count on fish for food.

Sea Thermal Energy

- Harnessing tidal and wave energy is not the only way to get electrical power from the ocean. Since 1979, the United States government has operated an Ocean Thermal Energy Conversion (OTEC) plant off the coast of Hawaii, where the temperature differential between surface and deeper water layers is converted into electrical energy.

Section 5

Ocean Pollution

Thermal Pollution

- Pesticides, oil, sludge, and trash are not the only harmful pollutants released into our oceans. Power plants can cause thermal pollution by releasing heated water into the sea. Thermal pollution may result in only a one- or two-degree temperature increase in the area near the heat source, but that can have profound effects on the ecology. Fish populations may migrate away from the affected area, and overgrowth of other organisms, or "algal blooms," may occur.

Close Quarters

- Seas that are surrounded by land are particularly vulnerable to damage from ocean pollution. The shores and adjacent waters of the Mediterranean, Baltic, and Adriatic Seas have been fouled with city sewage, factory waste, and fertilizer and pesticide runoff from farms. Their open waters have also been affected by dumping and oil spills.

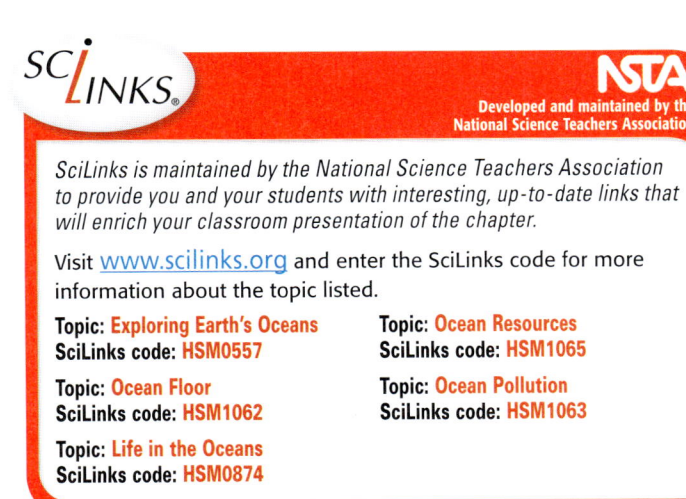

SciLINKS

NSTA
Developed and maintained by the National Science Teachers Association

SciLinks is maintained by the National Science Teachers Association to provide you and your students with interesting, up-to-date links that will enrich your classroom presentation of the chapter.

Visit www.scilinks.org and enter the SciLinks code for more information about the topic listed.

Topic: Exploring Earth's Oceans
SciLinks code: **HSM0557**

Topic: Ocean Floor
SciLinks code: **HSM1062**

Topic: Life in the Oceans
SciLinks code: **HSM0874**

Topic: Ocean Resources
SciLinks code: **HSM1065**

Topic: Ocean Pollution
SciLinks code: **HSM1063**

Overview

Tell students that this chapter will help them learn about the Earth's oceans. The chapter describes ocean characteristics, the ocean floor, life in the ocean, resources from the ocean, and ocean pollution.

Assessing Prior Knowledge

Students should be familiar with the following topics:

- plate tectonics
- the water cycle

Identifying Misconceptions

As students learn about the ocean as a resource, some of them may think that the oceans are a limitless resource. It may be helpful to review the definitions of *renewable resource* and *nonrenewable resource* before teaching about the resources from the ocean. Students may also think that vast bodies of water, such as the oceans, cannot be polluted. When teaching about ocean pollution, remind students that even though the oceans are very large, they are also susceptible to pollution.

2

Exploring the Oceans

About the PHOTO

Are two heads better than one? Although it may look like this reef lizardfish has two heads, it's actually swallowing another fish whole! Reef lizardfish are commonly found in the Western Pacific Ocean. Unlike most other types of lizardfish, the reef lizardfish prefers to rest on hard surfaces and is usually seen in pairs.

PRE-READING ACTIVITY

FOLDNOTES **Layered Book** Before you read the chapter, create the FoldNote entitled "Layered Book" described in the **Study Skills** section of the Appendix. Label the tabs of the layered book with "Characteristics of ocean water," "The ocean floor," "Ocean zones," and "Resources from the ocean." As you read the chapter, write information you learn about each category under the appropriate tab.

Standards Correlations

National Science Education Standards

The following codes indicate the National Science Education Standards that correlate to this chapter. The full text of the standards is at the front of the book.

Chapter Opener
UCP 2, 5; SAI 1; ST 2; SPSP 5

Section 1 Earth's Oceans
UCP 1, 2, 3; ES 1b, 1f, 1g, 1h, 1j, 2a

Section 2 The Ocean Floor
UCP 2, 3; SAI 1, 2; ST 2; SPSP 5; HNS 1, 3; ES 1b, 1c

Section 3 Life in the Ocean
UCP 1

Section 4 Resources from the Ocean
SAI 1; ST 2; SPSP 2, 4, 5; HNS 1

Section 5 Ocean Pollution
ST 2; SPSP 2, 4, 5; *LabBook:* UCP 2, 3; SAI 1; SPSP 2, 3, 4; HNS 1

START-UP ACTIVITY
MATERIALS

FOR EACH GROUP
- bowl, large
- cup, plastic, small, clear
- water

Teacher's Notes: A cylindrical cup (one that does not taper) is best for this activity.

Answers

1. Sample answer: The air inside the cup prevented the water below it from filling the space inside the cup.

2. Sample answer: The air in the cup keeps the water from filling the cup. Likewise, the air in an underwater research lab keeps water from coming through the hole in the bottom of the lab.

START-UP ACTIVITY

Exit Only?

To study what life underwater would be like, scientists sometimes live in underwater laboratories. How do these scientists enter and leave these labs? Believe it or not, the simplest way is through a hole in the lab's floor. You might think water would come in through the hole, but it doesn't. How is this possible? Do the following activity to find out.

Procedure

1. Fill a **large bowl** about two-thirds full of **water.**

2. Turn a **clear plastic cup** upside down.

3. Slowly guide the cup straight down into the water. Be careful not to guide the cup all the way to the bottom of the bowl. Also, be careful not to tip the cup.

4. Record your observations.

Analysis

1. How does the air inside the cup affect the water below the cup?

2. How do your findings relate to the hole in the bottom of an underwater research lab?

Chapter Lab
UCP 2, 3; SAI 1, 2; ST 2; SPSP 5; HNS 1

Chapter Review
UCP 1, 2, 3; SAI 2; ST 2; SPSP 5; HNS 1, 2; ES 1b, 1c, 1f, 1g, 1h, 1j

Science in Action
UCP 2; SAI 2; ST 2; SPSP 5; HNS 1, 2, 3

Chapter Starter Transparency
Use this transparency to help students begin thinking about exploring the oceans.

CHAPTER RESOURCES

Technology

📠 **Transparencies**   READING SKILLS
- Chapter Starter Transparency

💿 **Student Edition on CD-ROM**

💿 **Guided Reading Audio CD**
- English or Spanish

📹 **Classroom Videos**
- Brain Food Video Quiz

Workbooks

📖 **Science Puzzlers, Twisters & Teasers**
- Exploring the Oceans GENERAL

Focus

Overview

This section discusses how the oceans formed and how the global ocean is divided. It explores the properties of ocean water, including factors that affect salinity, temperature zones, and surface temperature changes.

 Bellringer

Show students a photo of Earth from space, and predict the percentage of land and water on Earth. (Liquid water covers 71% of Earth's surface.)

Motivate

ACTIVITY ——————— GENERAL

Ocean Size Discuss the divisions of the global ocean, and note that the volume of the Pacific Ocean is 724 million cubic kilometers. The volume of the Atlantic Ocean is about 322 million cubic kilometers, the volume of the Indian Ocean is about 292 million cubic kilometers, and the volume of the Arctic Ocean is about 12 million cubic kilometers. Have students fill graduated cylinders with water to demonstrate these ratios. (The Pacific Ocean would be 724 mL, the Atlantic Ocean would be 322 mL, the Indian Ocean would be 292 mL, and the Arctic Ocean would be 12 mL.) Visual

READING WARM-UP

Objectives
- List the major divisions of the global ocean.
- Describe the history of Earth's oceans.
- Identify the properties of ocean water.
- Describe the interactions between the ocean and the atmosphere.

Terms to Learn
salinity
water cycle

READING STRATEGY

Discussion Read this section silently. Write down questions that you have about this section. Discuss your questions in a small group.

Earth's Oceans

What makes Earth so different from Mars? What does Earth have that Mercury doesn't?

Earth stands out from the other planets in our solar system primarily for one reason—71% of the Earth's surface is covered with water. Most of Earth's water is found in the global ocean. The global ocean is divided by the continents into four main oceans. The divisions of the global ocean are shown in **Figure 1**. The ocean is a unique body of water that plays many parts in regulating Earth's environment.

Divisions of the Global Ocean

The largest ocean is the *Pacific Ocean*. It flows between Asia and the Americas. The volume of the *Atlantic Ocean*, the second-largest ocean, is about half the volume of the Pacific. The *Indian Ocean* is the third-largest ocean. The *Arctic Ocean* is the smallest ocean. This ocean is unique because much of its surface is covered by ice. Therefore, the Arctic Ocean has not been fully explored.

Figure 1 *The global ocean is divided by the continents into four main oceans.*

Arctic Ocean

Atlantic Ocean

Pacific Ocean

Indian Ocean

CHAPTER RESOURCES

Chapter Resource File
- Lesson Plan
- Directed Reading A **BASIC**
- Directed Reading B **SPECIAL NEEDS**

Technology

 Transparencies
- Bellringer
- Divisions of the Global Ocean

Is That a Fact!

The global ocean covers nearly 376 million square kilometers. The entire North American continent, by comparison, covers only a little more than 24 million square kilometers.

Figure 2 The History of Earth's Oceans

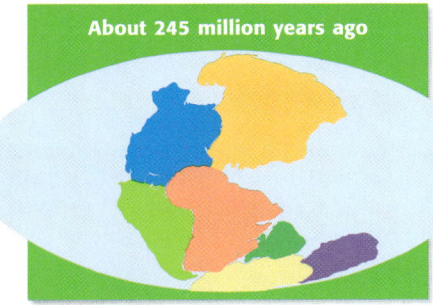

About 245 million years ago

The continents were one giant landmass called Pangaea. The oceans were one giant body of water called Panthalassa.

About 180 million years ago

As Pangaea broke apart, the North Atlantic Ocean and the Indian Ocean began to form.

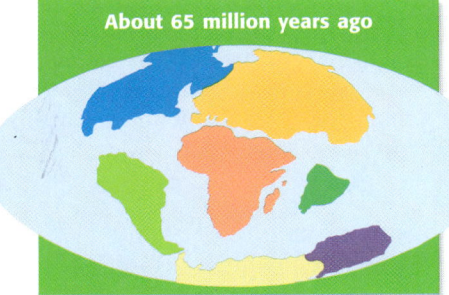

About 65 million years ago

The South Atlantic Ocean was much smaller than it is today.

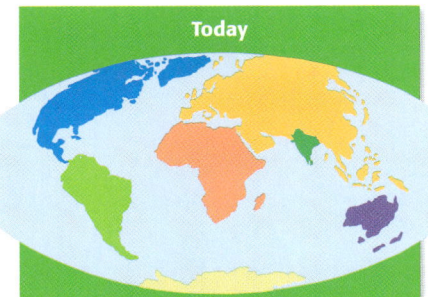

Today

The continents continue to move at a rate of 1 to 10 cm per year. The Pacific Ocean is getting smaller. However, the other oceans are growing.

How Did the Oceans Form?

About 4.5 billion years ago, Earth was a very different place. There were no oceans. Volcanoes spewed lava, ash, and gases all over the planet. The volcanic gases began to form Earth's atmosphere. Meanwhile, Earth was cooling. Sometime before 4 billion years ago, Earth cooled enough for water vapor to condense. This water began to fall as rain. The rain filled the deeper levels of Earth's surface, and the first oceans began to form.

The shape of the Earth's oceans has changed a lot over time. Much has been learned about the oceans' history. Some of this history is shown in **Figure 2**.

Reading Check How did the first oceans begin to form on Earth? (*See the Appendix for answers to Reading Checks.*)

MISCONCEPTION ALERT

Sea or Ocean? Students might find the terms *sea* and *ocean* confusing. In some cases, the words are interchangeable, but the terms can also mean different things. Parts of the global ocean that are partly surrounded by land are known as *seas*. The Mediterranean Sea is an example of a sea that is part of the global ocean. Seas that are completely landlocked, such as the Caspian Sea, are not part of the global ocean. Have students identify on a map landlocked seas and seas that are part of the global ocean.

Long Rainy Days When Earth cooled about 4 billion years ago, the rains that resulted probably lasted for thousands of years. But some scientists do not believe all the water on Earth came from condensation as Earth cooled. Instead, they argue that some of the water came from "cosmic rain"—comets that struck Earth in its early history. Encourage students to find out more about this debate.

MISCONCEPTION ALERT

The Composition of Ocean Water In **Figure 3,** students may notice that the percentages of some of the elements dissolved in ocean water are particularly low. This fact does not necessarily mean that these elements are not abundant in the ocean. Organisms, such as diatoms (phytoplankton) and coral, remove dissolved minerals containing some of these elements and use them to make hard body parts.

Ask students to note the two elements that are most abundant. (sodium and chlorine)

Ask students what these elements form when they are combined. (sodium chloride, or salt)

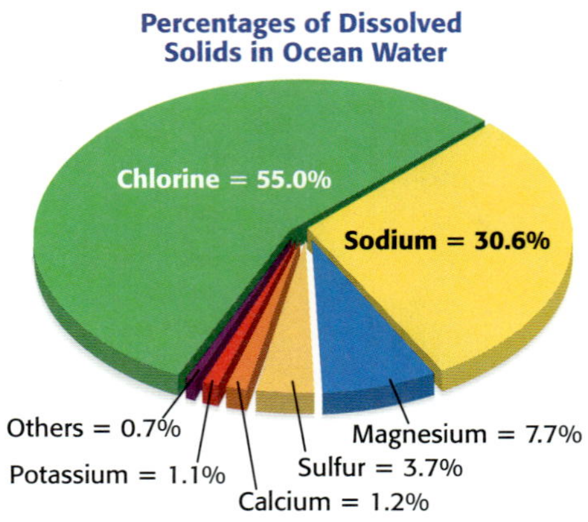

Percentages of Dissolved Solids in Ocean Water

Chlorine = 55.0%

Sodium = 30.6%

Others = 0.7%

Potassium = 1.1%

Sulfur = 3.7%

Calcium = 1.2%

Magnesium = 7.7%

Figure 3 *This pie graph shows the relative percentages of dissolved solids (by mass) in ocean water.*

salinity a measure of the amount of dissolved salts in a given amount of liquid

Characteristics of Ocean Water

You know that ocean water is different from the water that flows from your sink at home. For one thing, ocean water is not safe to drink. But there are other things that make ocean water special.

Ocean Water Is Salty

Have you ever swallowed water while swimming in the ocean? It tasted really salty, didn't it? Most of the salt in the ocean is the same kind of salt that we sprinkle on our food. This salt is called *sodium chloride.*

Salts have been added to the ocean for billions of years. As rivers and streams flow toward the oceans, they dissolve various minerals on land. The running water carries these dissolved minerals to the ocean. At the same time, water is *evaporating* from the ocean and is leaving the dissolved solids behind. The most abundant dissolved solid in the ocean is sodium chloride. This compound consists of the elements sodium, Na, and chlorine, Cl. **Figure 3** shows the relative amounts of the dissolved solids in ocean water.

Chock-Full of Solids

A measure of the amount of dissolved solids in a given amount of liquid is called **salinity.** Salinity is usually measured as grams of dissolved solids per kilogram of water. Think of it this way: 1 kg (1,000 g) of ocean water can be evaporated to 35 g of dissolved solids, on average. Therefore, if you evaporated 1 kg of ocean water, 965 g of fresh water would be removed and 35 g of solids would remain.

Climate Affects Salinity

Some parts of the ocean are saltier than others. Coastal water in places with hotter, drier climates typically has a higher salinity. Coastal water in cooler, more humid places typically has a lower salinity. One reason for this difference is that heat increases the evaporation rate. Evaporation removes water but leaves salts and other dissolved solids behind. Salinity levels are also lower in coastal areas that have a cooler, more humid climate because more fresh water from streams and rivers runs into the ocean in these areas.

✓ Reading Check Why does coastal water in places with hotter, drier climates typically have a higher salinity than coastal water in places with cooler, more humid climates?

CHAPTER RESOURCES

Technology

Transparencies
• Ocean Salinity

Answer to Reading Check

Coastal water in places with hotter, drier climates has a higher salinity because less fresh water flows into the ocean in drier areas and because heat increases the evaporation rate.

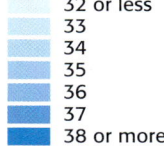

Proportion of salt per
1,000 parts of sea water

	32 or less
	33
	34
	35
	36
	37
	38 or more

Figure 4 *Salinity varies in different parts of the ocean because of variations in evaporation, circulation, and freshwater inflow.*

Water Movement Affects Salinity

Another factor that affects ocean salinity is water movement. Some parts of the ocean, such as bays, gulfs, and seas, move less than other parts. Parts of the open ocean that do not have currents running through them can also be slow moving. Slower-moving areas of water develop higher salinity. **Figure 4** shows salinity differences in different parts of the ocean.

Temperature Zones

The temperature of ocean water decreases as depth increases. However, this temperature change does not happen gradually from the ocean's surface to its bottom. Water in the ocean can be divided into three layers by temperature. As **Figure 5** shows, the temperature at the surface is much warmer than the average temperature of ocean water.

Figure 5	Temperature Zones in the Ocean

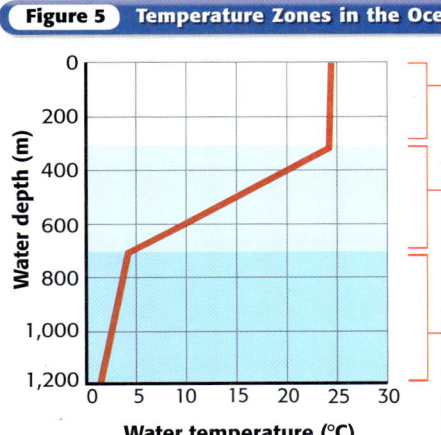

Surface zone The *surface zone* is the warm, top layer of ocean water. It can extend to 300 m below sea level. Sunlight heats the top 100 m of the surface zone. Surface currents mix the heated water with cooler water below.

Thermocline The *thermocline* is the second layer of ocean water. It can extend from 300 m below sea level to about 700 m below sea level. In the thermocline, temperature drops with increased depth faster than it does in the other two zones.

Deep zone The *deep zone* is the bottom layer that extends from the base of the thermocline to the bottom of the ocean. The temperature in this zone can range from 1°C to 3°C.

Making Models Using balloons and permanent markers, students can model how latitude affects ocean surface temperatures. Have students clearly label the poles and equator on their balloons and indicate surface water temperatures. Have them use colored pens to draw bands around their balloons. They should construct a key for their colors, correlating warmer temperatures with the colors closest to the equator. After they color in the shapes of the continents, students can present their models to the class. **LS Visual/Logical**

CONNECTION ACTIVITY
Geography ——————— GENERAL

Comparing the Oceans Encourage students to use atlases or globes to locate the four main oceans. Suggest that students draw a map of the oceans that indicates modern or ancient trade routes, and suggest that they illustrate the map with drawings of animals that are unique to each ocean. Students might also make a chart in which they compare all of the oceans by size, average temperature, depth, and other characteristics. **LS Visual**

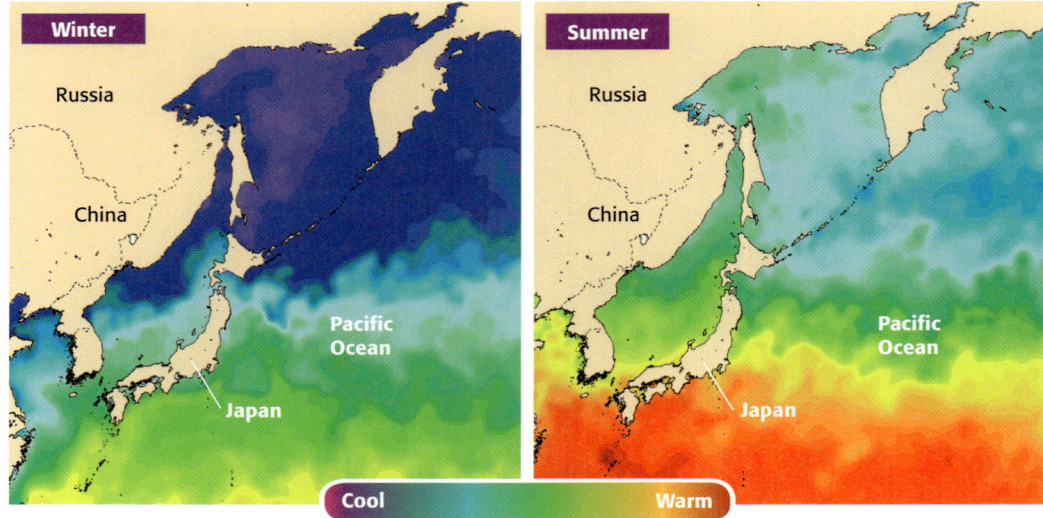

Figure 6 *These satellite images show that the surface temperatures in the northern Pacific Ocean change with the seasons.*

Surface Temperature Changes

If you live near the coast, you may know how different a swim in the ocean feels in December than it feels in July. Temperatures in the surface zone vary with latitude and the time of year. Surface temperatures range from 1°C near the poles to about 24°C near the equator. Parts of the ocean along the equator are warmer because they receive more direct sunlight per year than areas closer to the poles. However, both hemispheres receive more direct sunlight during their summer seasons. Therefore, the surface zone is heated more in the summer. **Figure 6** shows how surface-zone temperatures vary depending on the time of year.

✔ **Reading Check** Why are parts of the ocean along the equator warmer than those closer to the poles?

CONNECTION TO Geology

Submarine Volcanoes Geologists estimate that approximately 80% of the volcanic activity on Earth takes place on the ocean floor. Most of the volcanic activity occurs as magma slowly flows onto the ocean floor where tectonic plates pull away from each other. Other volcanic activity is the result of volcanoes that are located on the ocean floor. Both of these types of volcanoes are called *submarine volcanoes*. Submarine volcanoes behave differently than volcanoes on land do. Research how submarine volcanoes behave underwater. Then, create a model of a submarine volcano based on the information you find. **ACTIVITY**

Answer to Reading Check

Parts of the ocean along the equator are warmer because they receive more sunlight per year.

Is That a Fact!

Eighteen thousand years ago, much more of Earth's ocean water was frozen in glaciers and icecaps, and the Atlantic coast was miles farther out than it is today due to lower sea levels. Modern-day divers exploring the Chesapeake Bay found a mound of oyster shells—the remains of a long-ago picnic—40 m below present sea levels!

The Ocean and the Water Cycle

If you could sit on the moon and look down at Earth, what would you see? You would notice that Earth's surface is made up of three basic components—water, land, and clouds (air). All three are part of a process called the water cycle, as shown in **Figure 7.** The water cycle is the continuous movement of water from the ocean to the atmosphere to the land and back to the ocean. The ocean is an important part of the water cycle because nearly all of Earth's water is in the ocean.

water cycle the continuous movement of water from the ocean to the atmosphere to the land and back to the ocean

| Figure 7 | The Water Cycle |

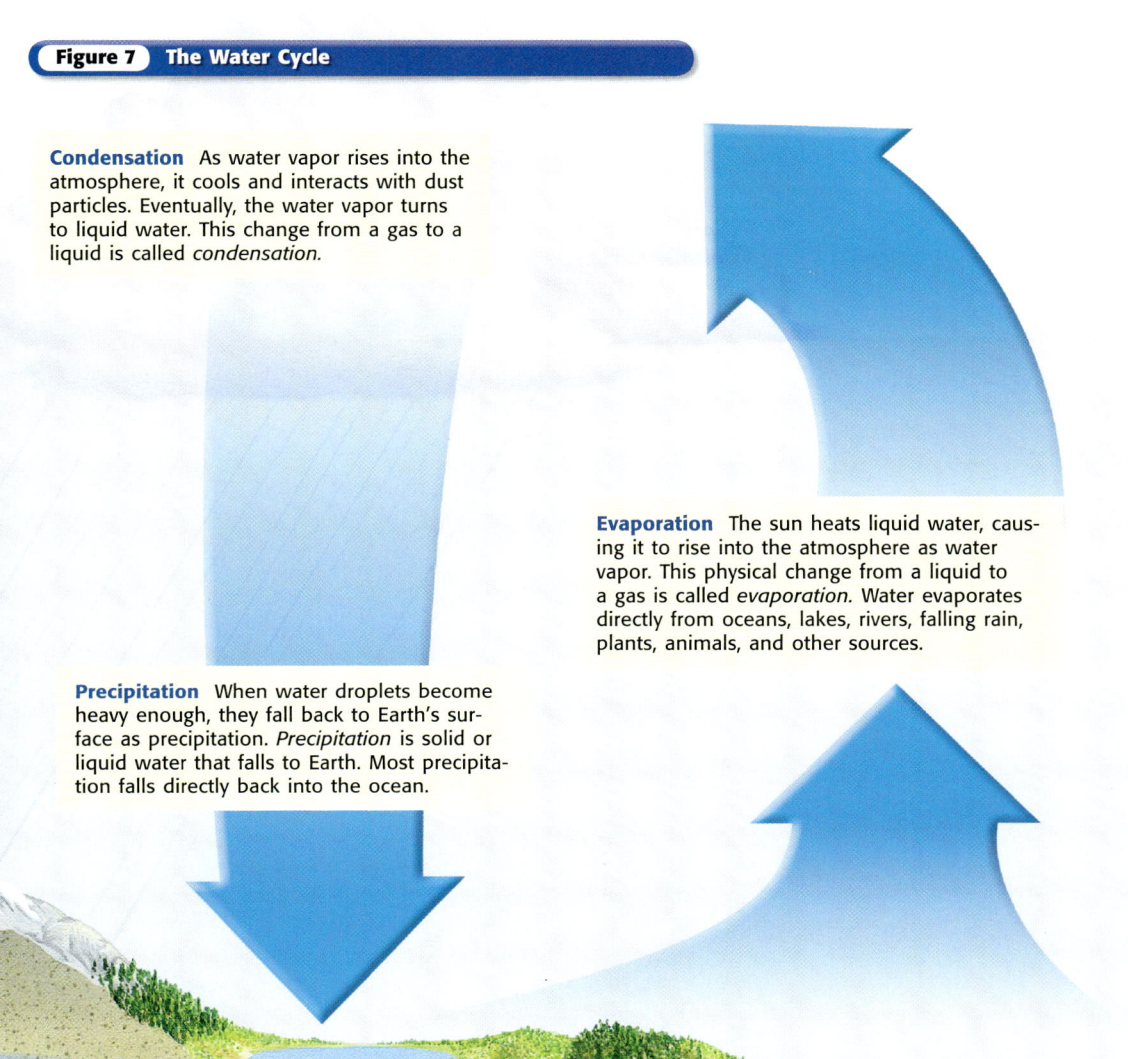

Condensation As water vapor rises into the atmosphere, it cools and interacts with dust particles. Eventually, the water vapor turns to liquid water. This change from a gas to a liquid is called *condensation.*

Evaporation The sun heats liquid water, causing it to rise into the atmosphere as water vapor. This physical change from a liquid to a gas is called *evaporation.* Water evaporates directly from oceans, lakes, rivers, falling rain, plants, animals, and other sources.

Precipitation When water droplets become heavy enough, they fall back to Earth's surface as precipitation. *Precipitation* is solid or liquid water that falls to Earth. Most precipitation falls directly back into the ocean.

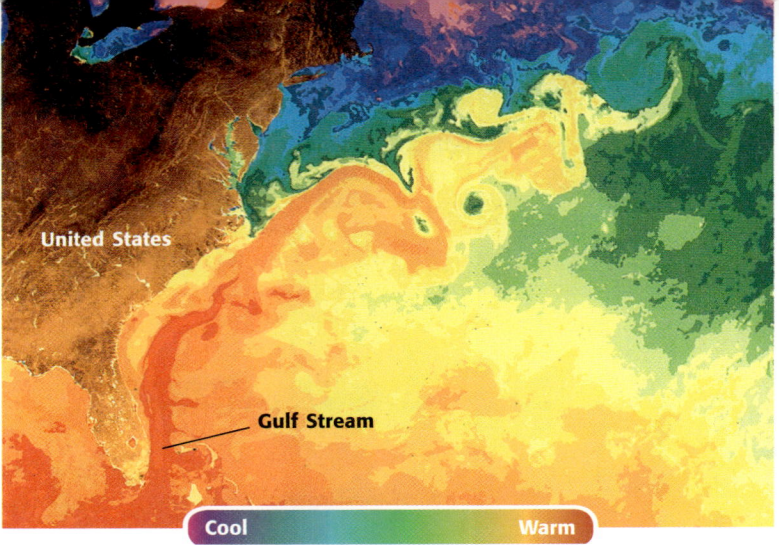

Figure 8 *This infrared satellite image shows the Gulf Stream moving warm water from lower latitudes to higher latitudes.*

United States

Gulf Stream

Cool _____ **Warm**

Close

Reteaching — BASIC

The Water Cycle List the steps of the water cycle on the board, and ask students to describe what happens at each step.
LS Verbal

Quiz — GENERAL

1. Which ocean is the largest? Which ocean is the smallest? (the Pacific Ocean; the Arctic Ocean)

2. How do scientists think the oceans are likely to change in the future? (They predict that the oceans will change in size and shape as the continents change position.)

Alternative Assessment — ADVANCED

PORTFOLIO **Ocean Timeline** Encourage students to examine **Figure 2.** Ask students to prepare a timeline detailing the history of Earth's oceans. Challenge students to predict how the oceans will change during the next 150 million years. Students can illustrate their timelines with drawings of each stage of Earth's history.
LS Visual

INTERNET ACTIVITY

For another activity related to this chapter, go to **go.hrw.com** and type in the keyword **HZ5OCEW.**

A Global Thermostat

The ocean plays an important part in keeping the Earth suitable for life. Perhaps the most important function of the ocean is to absorb and hold energy from sunlight. This function regulates temperatures in the atmosphere.

A Thermal Exchange

The ocean absorbs and releases thermal energy much more slowly than dry land does. If it were not for this property of the ocean, the air temperature on Earth could vary greatly from above 100°C during the day to below –100°C at night. This rapid exchange of thermal energy between the atmosphere and the Earth's surface would cause violent weather patterns. Life as you know it could not exist under these conditions.

✓ **Reading Check** How would the air temperature on land be different if the ocean did not release thermal energy so slowly?

Have Heat, Will Travel

The ocean also regulates temperatures at different locations of the Earth. At the equator, the sun's rays are more direct than at the poles. As a result, the waters there are warmer than waters at higher latitudes. However, currents in the ocean move water and the energy it contains. Part of this movement is shown in **Figure 8.** This circulation of warm water causes some coastal lands to have warmer climates than they would have without the currents. The British Isles, for example, have a warmer climate than most regions at the same latitude. This warmer climate is due to the warm water of the Gulf Stream.

Answer to Reading Check
If the ocean did not release thermal energy so slowly, the air temperature on land would vary greatly from above 100°C during the day to below -100°C at night.

MISCONCEPTION ALERT

Sea Level Students may think that sea level is the same worldwide. Tides and winds alter the ocean's depth constantly. Pacific Ocean trade winds blow westward, causing the ocean level to be about a half meter higher on the western side of the Pacific. Sea level is higher at the equator than at the poles because the equatorial waters expand and the centrifugal force of Earth's rotation causes the middle of the planet to bulge.

SECTION Review

Summary

- The global ocean is divided by the continents into four main oceans: Pacific Ocean, Atlantic Ocean, Indian Ocean, and Arctic Ocean.
- The four oceans as we know them today formed within the last 300 million years.
- Salts have been added to the ocean for billions of years. Salinity is a measure of the amount of dissolved salts in a given weight or mass of liquid.

- The three temperature zones of ocean water are the surface zone, the thermocline, and the deep zone.
- The water cycle is the continuous movement of water from the ocean to the atmosphere to the land and back to the ocean. The ocean plays the largest role in the water cycle.
- The ocean stabilizes Earth's weather conditions by absorbing and holding thermal energy.

Using Key Terms

1. In your own words, write a definition for each of the following terms: *salinity* and *water cycle*.

Understanding Key Ideas

2. The top layer of ocean water that extends to 300 m below sea level is called the
   - a. deep zone.
   - b. surface zone.
   - c. Gulf Stream.
   - d. thermocline.
3. Name the major divisions of the global ocean.
4. Explain how Earth's first oceans formed.
5. Why is the ocean an important part of the water cycle?
6. Between which two steps of the water cycle does the ocean fit?

Critical Thinking

7. **Making Inferences** Describe how the ocean plays a role in stabilizing Earth's weather conditions.
8. **Identifying Relationships** List one factor that affects salinity in the ocean and one factor that affects ocean temperatures. Explain how each factor affects salinity or temperature.

Interpreting Graphics

Use the image below to answer the questions that follow.

9. At which stage would solid or liquid water fall to the Earth?
10. At which stage would the sun's energy cause liquid to rise into the atmosphere as water vapor?

Developed and maintained by the National Science Teachers Association

For a variety of links related to this chapter, go to www.scilinks.org

Topic: Exploring Earth's Oceans
SciLinks code: HSM0557

SCIENCE HUMOR

Q: Why is the ocean salty?

A: because fish don't like pepper

CHAPTER RESOURCES

Chapter Resource File

- Section Quiz GENERAL
- Section Review GENERAL
- Vocabulary and Section Summary GENERAL

Answers to Section Review

1. Sample answer: Salinity is a measure of the amount of dissolved salts in a given amount of liquid. The water cycle is the continuous movement of water from the ocean to the atmosphere to the land and back to the ocean.
2. b
3. The major divisions of the global ocean are the Pacific Ocean, the Atlantic Ocean, the Indian Ocean, and the Arctic Ocean.
4. Sample answer: Sometime before 4 billion years ago, Earth cooled enough for water vapor to condense. This water began to fall as rain. The rain filled the deeper levels of Earth's surface, and the first oceans began to form.
5. Sample answer: The ocean is an important part of the water cycle because nearly all of Earth's water is found in the ocean and the ocean absorbs the majority of the solar radiation that reaches the Earth.
6. The ocean fits between precipitation and evaporation in the water cycle.
7. The ocean helps to stabilize Earth's weather conditions because it absorbs and releases thermal energy from sunlight slowly. This function regulates temperatures in the atmosphere.
8. Sample answer: Evaporation affects salinity in the ocean, and depth affects the temperature of the ocean. When water evaporates from the ocean, dissolved solids are left behind. Therefore, high rates of evaporation leave the ocean saltier. The temperature of ocean water decreases as depth increases. Therefore, the temperature at the surface is much warmer than the average temperature of the ocean.
9. precipitation
10. evaporation

SECTION
2

Focus

Overview

This section discusses how technology has facilitated exploration of the ocean floor and the methods used to survey the ocean floor, including sonar and satellites. This section also discusses the regional divisions of the ocean floor and the geographical features of each division.

🔔 Bellringer

Have students pretend that they have walked off the edge of North America and into the depths of the Atlantic Ocean. As they walk along the ocean floor toward Europe, what would they see? Have students make a drawing of the ocean floor that they would see along the way.

Motivate

Discussion ─────── GENERAL

The Ocean Floor It has been said that scientists know more about the surface of the moon than about the ocean floor. Most of what scientists know about the ocean floor comes from sonar readings and sample dredging. Ask students to think about why the ocean is so difficult to study and what kinds of technology would help scientists learn more about the deep-ocean floor. **LS Verbal**

READING WARM-UP

Objectives

- Describe technologies for studying the ocean floor.
- Identify the two major regions of the ocean floor.
- Classify subdivisions and features of the two major regions of the ocean floor.

Terms to Learn

continental shelf	rift valley
continental slope	seamount
continental rise	ocean trench
abyssal plain	
mid-ocean ridge	

READING STRATEGY

Reading Organizer As you read this section, create an outline of the section. Use the headings from the section in your outline.

The Ocean Floor

What lies at the bottom of the ocean? How deep is the ocean?

These questions were once unanswerable. By using new technology, scientists have learned a lot about the ocean floor. Scientists have discovered landforms on the ocean floor and have measured depths for almost the entire ocean floor.

Studying the Ocean Floor

Sending people into deep water to study the ocean floor can be risky. Fortunately, there are other ways to study the deep ocean. These ways include surveying from the ocean surface and from high above in space.

Seeing by Sonar

Sonar stands for *sound navigation and ranging*. This technology is based on the echo-ranging behavior of bats. Scientists use sonar to determine the ocean's depth by sending sound pulses from a ship down into the ocean. The sound moves through the water, bounces off the ocean floor, and returns to the ship. The deeper the water is, the longer the round trip takes. Scientists then calculate the depth by multiplying half the travel time by the speed of sound in water (about 1,500 m/s). This process is shown in **Figure 1.**

Figure 1 Ocean Floor Mapping with Sonar

Scientists use sonar signals to make a *bathymetric profile,* which is a map of the ocean floor that shows the ocean's depth.

CHAPTER RESOURCES

Chapter Resource File

- Lesson Plan
- Directed Reading A **BASIC**
- Directed Reading B **SPECIAL NEEDS**

Technology

- Transparencies
  - Bellringer
  - How Sonar Works

 **WEIRD SCIENCE**

Although many people know that whales and dolphins communicate by sound, few people are aware that shrimp do the same thing. To locate food sources and other shrimp, they emit a sound similar to the sound of bacon frying!

Oceanography via Satellite

In the 1970s, scientists began studying Earth from satellites in orbit around the Earth. In 1978, scientists launched the satellite *Seasat*. This satellite focused on the ocean, sending images back to Earth that allowed scientists to measure the direction and speed of ocean currents.

Studying the Ocean with *Geosat*

Geosat, once a top-secret military satellite, has been used to measure slight changes in the height of the ocean's surface. Different underwater features, such as mountains and trenches, affect the height of the water above them. Scientists measure the different heights of the ocean surface and use the measurements to make detailed maps of the ocean floor. Maps made using satellite measurements, such as the map in **Figure 2,** can cover much more territory than maps made using ship-based sonar readings.

✔ **Reading Check** How do scientists use satellites to make detailed maps of the ocean floor? (*See the Appendix for answers to Reading Checks.*)

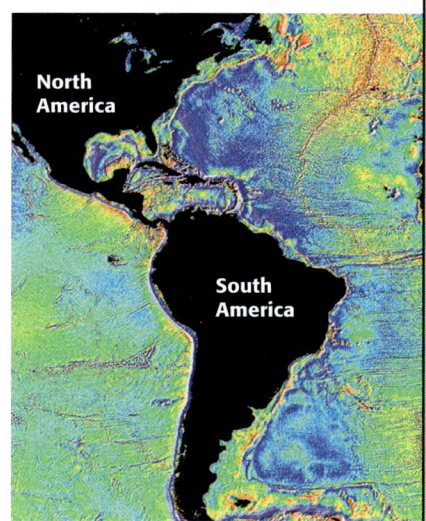

Figure 2 *This map was generated by satellite measurements of different heights of the ocean surface.*

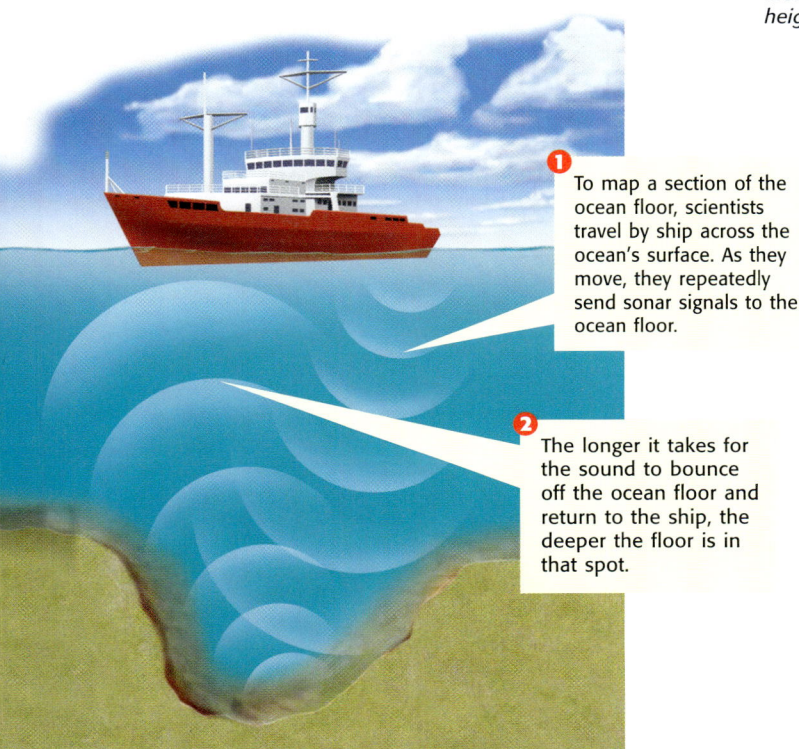

❶ To map a section of the ocean floor, scientists travel by ship across the ocean's surface. As they move, they repeatedly send sonar signals to the ocean floor.

❷ The longer it takes for the sound to bounce off the ocean floor and return to the ship, the deeper the floor is in that spot.

Answer to Reading Check

Satellite photos from *Seasat* send images of the ocean back to Earth. These images allow scientists to measure the direction and speed of ocean currents. Satellite photos and information from *Geosat* have been used to measure slight changes in the height of the ocean's surface.

Water Pressure To show how water pressure changes with depth, punch three holes in the side of a milk carton: one near the top, one halfway down the side, and one near the bottom. Put one piece of tape over all three holes, and fill the carton with water. Remove the tape quickly. Have students observe the streams of water and explain what they see. (The water stream at the bottom of the carton will shoot out the farthest and with the greatest force. The reason is that the water at the top of the carton exerted pressure on the water at the bottom of the carton.) **LS Visual/Kinesthetic**

Homework — ADVANCED

Concept Mapping Remind students that volcanic seamounts that rise above the ocean surface become volcanic islands. The Hawaiian Islands formed this way. Have students research other ways islands form and then prepare a concept map of the different ways that islands form. They should find that some islands form by the growth of coral, some form by deposition (barrier islands), and some are continental islands (for example, Great Britain and Madagascar). **LS Visual**

continental shelf the gently sloping section of the continental margin located between the shoreline and the continental slope

continental slope the steeply inclined section of the continental margin located between the continental rise and the continental shelf

continental rise the gently sloping section of the continental margin located between the continental slope and the abyssal plain

abyssal plain a large, flat, almost level area of the deep-ocean basin

Revealing the Ocean Floor

Can you imagine being an explorer assigned to map uncharted areas on the planet? You might think that there are not many uncharted areas left because most of the land has already been explored. But what about the bottom of the ocean?

The ocean floor is not a flat surface. If you could go to the bottom of the ocean, you would see a number of impressive features. You would see the world's longest mountain chain, which is about 64,000 km (40,000 mi) long as well as canyons deeper than the Grand Canyon. And because it is underwater and some areas are so deep, much of the ocean floor is still not completely explored.

✓ **Reading Check** How long is the longest mountain chain in the world? Where is it located?

Figure 3   The Ocean Floor

The **continental shelf** begins at the shoreline and slopes gently toward the open ocean. It continues until the ocean floor begins to slope more steeply downward. The depth of the continental shelf can reach 200 m.

The **continental slope** begins at the edge of the continental shelf. It continues down to the flattest part of the ocean floor. The depth of the continental slope ranges from about 200 m to about 4,000 m.

The **continental rise,** which is the base of the continental slope, is made of large piles of sediment. The boundary between the continental margin and the deep-ocean basin lies underneath the continental rise.

The **abyssal plain** is the broad, flat part of the deep-ocean basin. It is covered by mud and the remains of tiny marine organisms. The average depth of the abyssal plain is about 4,000 m.

Answer to Reading Check
64,000 km; on the ocean floor

SCIENCE HUMOR

Q: What lies on the bottom of the ocean and trembles?

A: a nervous wreck

Regions of the Ocean Floor

If you journeyed to the ocean floor, you would first notice two major regions. The *continental margin* is made of continental crust, and the *deep-ocean basin* is made of oceanic crust. Imagine that the ocean is a giant swimming pool. The continental margin is the shallow end of the pool, and the deep-ocean basin is the deep end of the pool. The figure below shows how these two regions are subdivided.

Underwater Real Estate

As you can see in **Figure 3** below, the continental margin is subdivided into the continental shelf, the continental slope, and the continental rise. These divisions are based on depth and changes in slope. The deep-ocean basin consists of the abyssal (uh BIS uhl) plain, mid-ocean ridges, rift valleys, and ocean trenches. All of these features form near the boundaries of Earth's *tectonic plates*. On parts of the deep-ocean basin that are not near plate boundaries, there are thousands of seamounts. Seamounts are submerged volcanic mountains on the ocean floor.

✓ Reading Check What are the subdivisions of the continental margin?

mid-ocean ridge a long, undersea mountain chain that forms along the floor of the major oceans

rift valley a long, narrow valley that forms as tectonic plates separate

seamount a submerged mountain on the ocean floor that is at least 1,000 m high and that has a volcanic origin

ocean trench a steep, long depression in the deep-sea floor that runs parallel to a chain of volcanic islands or a continental margin

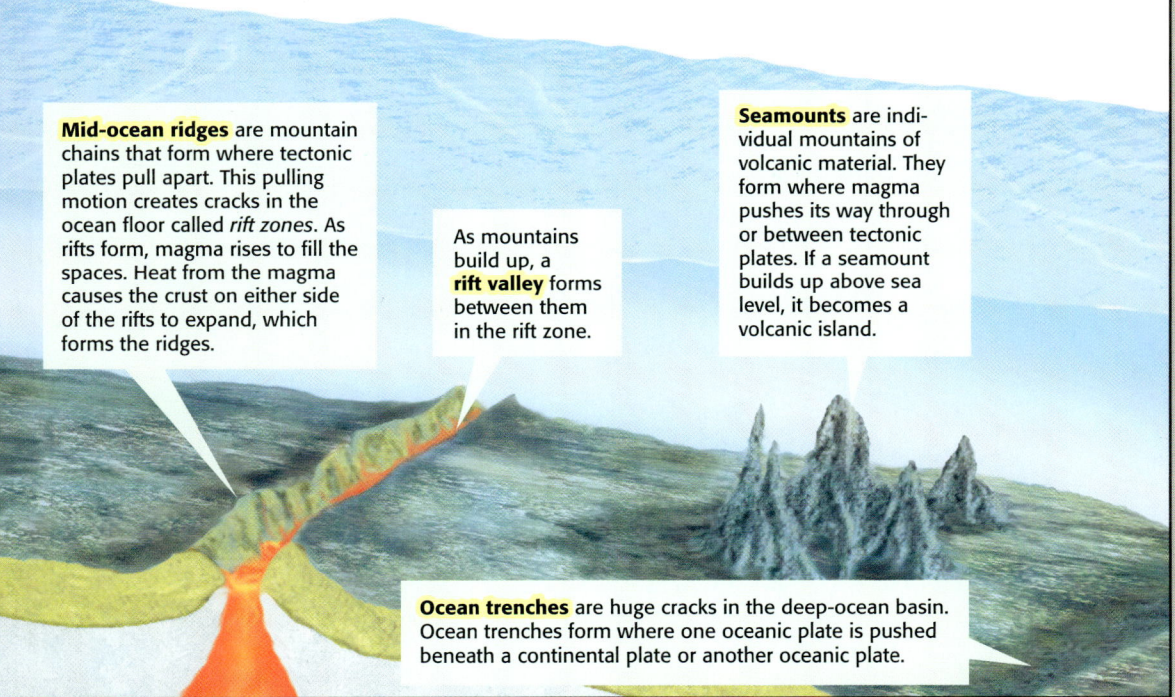

Mid-ocean ridges are mountain chains that form where tectonic plates pull apart. This pulling motion creates cracks in the ocean floor called *rift zones*. As rifts form, magma rises to fill the spaces. Heat from the magma causes the crust on either side of the rifts to expand, which forms the ridges.

As mountains build up, a **rift valley** forms between them in the rift zone.

Seamounts are individual mountains of volcanic material. They form where magma pushes its way through or between tectonic plates. If a seamount builds up above sea level, it becomes a volcanic island.

Ocean trenches are huge cracks in the deep-ocean basin. Ocean trenches form where one oceanic plate is pushed beneath a continental plate or another oceanic plate.

Answer to Reading Check

continental shelf, continental slope, and continental rise

CHAPTER RESOURCES

Technology

Transparencies
• Revealing the Ocean Floor: A
• Revealing the Ocean Floor: B

Using the Figure — GENERAL

Mid-Ocean Ridges Draw students' attention to the mid-ocean ridges in **Figure 3.** Below the rift zones that characterize mid-ocean ridges, magma rises from beneath the crust and erupts as lava. The lava cools when it enters the water and forms new oceanic crust. Point out that ocean trenches formed by the subduction of plates are some of the deepest places on Earth and often support a diversity of life. **LS Verbal**

CONNECTION ACTiViTY
Art ——————— GENERAL

Illustrating the Ocean Floor Draw students' attention to **Figure 3,** and have them draw, label, and color their own illustration of the depth zones of the ocean floor. Point out the canyon in the continental slope. This canyon is a submarine canyon. Most of the sediment that makes up the continental rise travels down from the continental shelf through submarine canyons. Be sure that students have divided the continental margin into the continental shelf, continental slope, and continental rise. They should identify the features of the deep-ocean basin as mid-ocean ridges, seamounts, rift valleys, and ocean trenches. In addition, students can indicate the temperature of the ocean water at each depth by using different colors. **LS Visual** **English Language Learners**

Reteaching ——— BASIC

The Ocean Floor Have students review the regions and features of the ocean floor. Ask students to choose any two regions or features and to describe them in their own words. If describing features, students should include how the features form.

LS Verbal

Quiz ——— GENERAL

1. Which features of the abyssal plain form at the boundaries of tectonic plates? (mid-ocean ridges, rift valleys, and ocean trenches)

2. How is the depth of the ocean measured? (It is measured using sonar. Scientists calculate the depth by multiplying half the time a sound wave takes to hit the ocean floor and return to the surface by the speed of sound in water.)

Alternative Assessment ——— GENERAL

Modeling the Ocean Floor
Encourage students to imagine that they are oceanographers on a deep-sea mission aboard a piloted vessel. Have them work in groups to make a model of the ocean floor, including all the features from this section. The model can be a cross section similar to **Figure 3** or a view from above.
LS Visual/Verbal   English Language Learners

CONNECTION TO Social Studies

The JASON Project The JASON project, started by oceanographer Dr. Robert Ballard, allows students and teachers to take part in virtual field trips to some of the most exotic locations on Earth. Using satellite links and the Internet, students around the world have participated in scientific expeditions to places such as the Galápagos Islands, the Sea of Cortez, and deep-sea hydrothermal vents. Using the Internet, research where the JASON project is headed to next!

Exploring the Ocean with Underwater Vessels

Just as astronauts explore space with rockets, scientists explore the oceans with underwater vessels. These vessels contain the air that the explorers need to breathe and all of the scientific instruments that the explorers need to study the oceans.

Piloted Vessels: *Alvin* and *Deep Flight*

One research vessel used to travel to the deep ocean is called *Alvin*. *Alvin* is 7 m long and can reach some of the deepest parts of the ocean. Scientists have used *Alvin* for many underwater missions, including searches for sunken ships, the recovery of a lost hydrogen bomb, and explorations of the sea floor. In 1977, scientists aboard *Alvin* discovered an oasis of life around hydrothermal vents near the Galápagos Islands. Ecosystems near hydrothermal vents are unique because some organisms living around the vent do not rely on photosynthesis for energy. Instead, these organisms rely on chemicals in the water as their source of energy.

Another modern vessel that scientists use to explore the deep ocean is an underwater airplane called *Deep Flight*. This vessel, shown in **Figure 4,** moves through the water in much the same way that an airplane moves through the air. Future models of *Deep Flight* will be designed to transport pilots to the deepest parts of the ocean, which are more than 11,000 m deep.

✓ **Reading Check** Why is the ecosystem discovered by *Alvin* unique?

Figure 4 *Like the Wright brothers' first successful airplane, Deep Flight sets the stage for a bright future— this time in underwater "flight."*

Answer to Reading Check

It is unique because some of the organisms living around the vent do not rely on photosynthesis for energy.

Is That a Fact!

The same explorer who led the first voyage around the world also attempted to determine the depth of the ocean. In 1520, Ferdinand Magellan weighted a 370 m rope with lead and lowered it into the ocean. But, his rope was not long enough to reach the ocean floor! The first successful measurement was made in 1773. Using Magellan's techniques, explorers found that the depth of the ocean near Norway is about 1,250 m.

Robotic Vessels: *JASON II* and *Medea*

Exploring the deep ocean by using piloted vessels is expensive and can be very dangerous. For these reasons, scientists use robotic vessels to explore the ocean. One interesting robot team consists of *JASON II* and *Medea*. These robots are designed to withstand pressures much greater than those found in the deepest parts of the ocean. *JASON II* is "flown" by a pilot at the surface and is used to explore the ocean floor. *Medea* is attached to *JASON II* with a tether and explores above the sea floor. In the future, unpiloted "drone" robots shaped like fish may be used. Another robot under development uses the ocean's thermal energy for power. These robots could explore the ocean for years and send data to scientists at the surface.

SECTION Review

Summary

- Scientists study the ocean floor from the surface using sonar and satellites.
- The ocean floor is divided into two regions—the continental margin and the deep-ocean basin.
- The continental margin consists of the continental shelf, the continental slope, and the continental rise.
- The deep-ocean basin consists of the abyssal plain, mid-ocean ridges, rift valleys, seamounts, and ocean trenches.
- Scientists explore the ocean from below the surface by using piloted vessels and robotic vessels.

Using Key Terms

For each pair of terms, explain how the meanings of the terms differ.

1. *continental shelf* and *continental slope*
2. *abyssal plain* and *ocean trench*
3. *mid-ocean ridge* and *seamount*

Understanding Key Ideas

4. Sonar is a technology based on the
   a. *Geosat* satellite.
   b. surface currents in the ocean.
   c. zones of the ocean floor.
   d. echo-ranging behavior of bats.

5. List the two major regions of the ocean floor.

6. Describe the subdivisions of the continental margin.

7. List three technologies for studying the ocean floor, and explain how they are used.

8. List three underwater missions that *Alvin* has been used for.

9. Explain how *Jason II* and *Medea* are used to explore the ocean.

10. Describe how a bathymetric profile is made.

Math Skills

11. Air pressure at sea level is 1 atmosphere (atm). Underwater, pressure increases by 1 atm every 10 m of depth. For example, at a depth of 10 m, water pressure is 2 atm. What is the pressure at 100 m?

Critical Thinking

12. **Making Comparisons** How is exploring the oceans similar to exploring space?

13. **Applying Concepts** Is the ocean floor a flat surface? Explain your answer.

Answers to Section Review

1. Sample answer: The continental shelf is the gently sloping section of the continental margin. The continental slope is the steeply inclined section of the continental margin.

2. Sample answer: The abyssal plain is a large, flat, almost-level area of the deep-ocean basin. An ocean trench is a steep, long depression in the deep-sea floor.

CHAPTER RESOURCES

Chapter Resource File

- Section Quiz **GENERAL**
- Section Review **GENERAL**
- Vocabulary and Section Summary **GENERAL**

3. Sample answer: A midocean ridge is a long, undersea mountain chain that forms along the floor of the major oceans. A seamount is a submerged mountain on the ocean floor.

4. d

5. continental margin and deep-ocean basin

6. The continental shelf is the gently sloping section located between the shoreline and the continental slope. The continental slope is the steeply inclined section of the continental margin. The continental rise is the gently sloping section located between the continental slope and the abyssal plain.

7. Sample answer: sonar, *Geosat,* and underwater vessels; Sonar is used to determine the ocean's depth by sending sound pulses from a ship down into the ocean. *Geosat* is a satellite that is used to measure the different heights of the ocean's surface, which can help detect mountains and trenches on the ocean floor. Underwater vessels (both piloted and robotic) are used to travel to the deepest parts of the ocean.

8. searches for sunken ships, the recovery of a lost hydrogen bomb, and explorations of the sea floor

9. Sample answer: *Jason II* is "flown" by a pilot at the surface and is used to explore the ocean floor. *Medea* is attached to *Jason II* with a tether and explores above the sea floor.

10. Scientists use sonar signals to make a bathymetric profile.

11. 11 atm

12. Sample answer: Exploring the oceans is similar to exploring space because it can be dangerous to explore both and much is still unknown about the ocean and about space.

13. Sample answer: no; The ocean floor is not a flat surface because it has many features, including the world's longest mountain chain.

SECTION
3

Focus

Overview

This section introduces a system for classifying marine organisms based on where they live and how they move. Students also learn to describe ecological zones of the ocean and give examples of organisms inhabiting each zone.

Bellringer

Before they read this section, have students imagine they are marine biologists who must classify marine life into three groups. Challenge them to identify the criteria they would use in their classification systems.

Motivate

Group ACTiViTY — GENERAL

Classifying Divide the class into small groups. Ask them to classify as many items in the classroom as they can based on the following categories:

• height at which the items are located

• ways the items are used

 Visual/Verbal

READING WARM-UP

Objectives

● Identify the three groups of marine life.

● Describe the two main ocean environments.

● Identify the ecological zones of the benthic and pelagic environments.

Terms to Learn

plankton
nekton
benthos
benthic environment
pelagic environment

READING STRATEGY

Mnemonics As you read this section, create a mnemonic device to help you remember the ecological zones of the ocean.

Life in the Ocean

In which part of the ocean does an octopus live? And where do dolphins spend most of their time?

Just as armadillos and birds occupy very different places on Earth, octopuses and dolphins live in very different parts of the ocean. Trying to study life in the oceans can be a challenge for scientists. The oceans are so large that many forms of marine life have not been discovered, and there are many more organisms that scientists know little about. To make things easier, scientists classify marine organisms into three main groups.

The Three Groups of Marine Life

The three main groups of marine life, as shown in **Figure 1,** are plankton, nekton, and benthos. Marine organisms are placed into one of these three groups according to where they live and how they move.

Organisms that float or drift freely near the ocean's surface are called **plankton.** Most plankton are microscopic. Plankton are divided into two groups—those that are plant-like (*phytoplankton*) and those that are animal-like (*zooplankton*). Organisms that swim actively in the open ocean are called **nekton.** Types of nekton include mammals, such as whales, dolphins, and sea lions, as well as many varieties of fish. **Benthos** are organisms that live on or in the ocean floor. There are many types of benthos, such as crabs, starfish, worms, coral, sponges, seaweed, and clams.

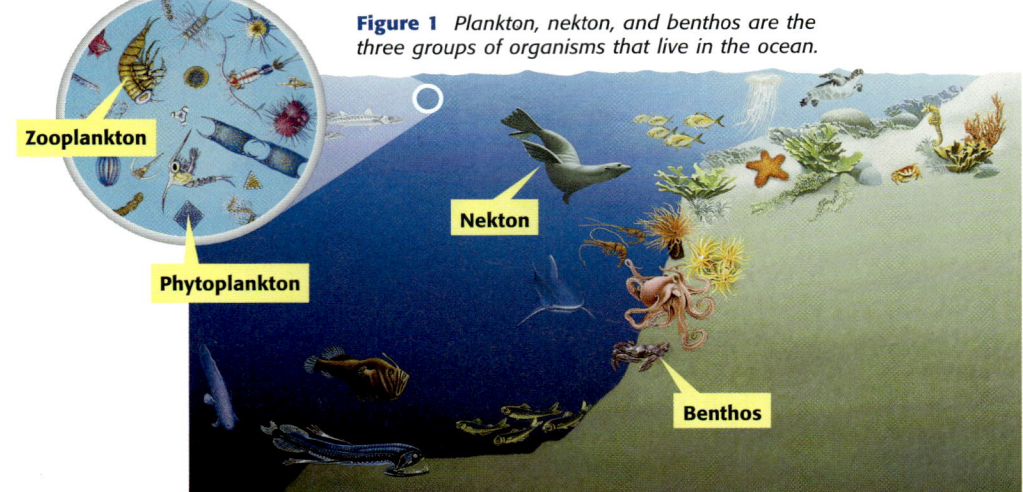

Figure 1 *Plankton, nekton, and benthos are the three groups of organisms that live in the ocean.*

Zooplankton

Phytoplankton

Nekton

Benthos

CHAPTER RESOURCES

Chapter Resource File

• Lesson Plan
• Directed Reading A BASIC
• Directed Reading B SPECIAL NEEDS

Technology

Transparencies
• Bellringer
• The Three Groups of Marine Life

Is That a Fact!

The word *plankton* comes from the Greek *planktos,* meaning "wandering." Because plankton float at or near the ocean's surface, they "wander" with the ocean currents. Interestingly, tiny plankton are the sole sustenance of two of the largest marine animals—the blue whale and the basking shark. The word *planets* is also derived from the same root and refers to the observation that planets appear to wander among the stars.

The Benthic Environment

In addition to being divided into zones based on depth, the ocean floor is divided into ecological zones based on where different types of benthos live. These zones are grouped into one major marine environment—the benthic environment. The **benthic environment,** or bottom environment, is the region near the ocean floor and all the organisms that live on or in it.

The Intertidal Zone

The shallowest benthic zone, called the *intertidal zone,* is located between the low-tide and high-tide limits. Twice a day, the intertidal zone changes. As the tide flows in, the zone is covered with ocean water. Then, as the tide flows out, the intertidal zone is exposed to the air and sun.

Because of the change in tides, intertidal organisms must be able to live both underwater and on exposed land. Some organisms, such as the sea anemones and starfish shown in **Figure 2,** attach themselves to rocks and reefs to avoid being washed out to sea during low tide. Other organisms, such as clams, oysters, barnacles, and crabs, have tough shells that give them protection against strong waves during high tide and against harsh sunlight during low tide. Some animals can burrow in sand or between rocks to avoid harsh conditions. Plants also protect themselves from being washed away by strong waves. Plants such as seaweed have strong *holdfasts* (rootlike structures) that allow them to grow in this zone.

✔ Reading Check How do clams and oysters survive in the intertidal zone during high tide and low tide? (*See the Appendix for answers to Reading Checks.*)

Intertidal zone

Figure 2 *Organisms such as sea anemones and starfish attach themselves to rocks and reefs. These organisms must be able to survive both wet and dry conditions.*

plankton the mass of mostly microscopic organisms that float or drift freely in freshwater and marine environments

nekton all organisms that swim actively in open water, independent of currents

benthos the organisms that live at the bottom of the sea or ocean

benthic environment the region near the bottom of a pond, lake, or ocean

SCIENTISTS AT ODDS

The Azoic Theory Before the late 1870s, scientists widely believed the "azoic theory" of James Forbes and Alexander Agassiz. This theory argued that no life existed below shallow depths in the oceans. Sir Wyville Thomson disputed this theory, and in 1872, he embarked on a three-and-a-half year voyage to prove his point. The HMS *Challenger* voyage, which established oceanography as a modern science, collected ocean-floor samples from deeper than 8,000 m. Thomson's evidence disproved the azoic theory and greatly expanded our knowledge of the world's oceans. He published his results in *The Depths of the Sea,* the first general textbook on oceanography.

Teach, continued

Group ACTIVITY — GENERAL

Ocean Zones and Organisms
Divide the class into groups of five, and provide each group with poster board and markers. Ask groups to draw a cross-sectional illustration of the ocean and label the following features: benthic environment, intertidal zone, sublittoral zone, bathyal zone, abyssal zone, and hadal zone.

Have each group member illustrate the organisms found in one of the five zones. (Each group should illustrate all of the five zones.) Instruct each group to elect a spokesperson to present the poster to the class.
LS Visual/Intrapersonal

Homework ——— GENERAL

Benthic Research
Have students select one of the benthic zones to investigate further. Ask them to focus on the adaptations that organisms living there have developed that enable them to exist in that zone. Students can present their findings in a concept map, poster, or comic book.
English Language Learners
LS Interpersonal

Figure 3 *Corals, like many other types of organisms, can live in both the sublittoral zone and the intertidal zone. However, they are more common in the sublittoral zone.*

Sublittoral zone

The Sublittoral Zone

The *sublittoral zone* begins where the intertidal zone ends, at the low-tide limit, and extends to the edge of the continental shelf. This zone of the benthic environment is more stable than the intertidal zone. The temperature, water pressure, and amount of sunlight remain fairly constant in the sublittoral zone. Sublittoral organisms, such as corals, shown in **Figure 3,** do not have to cope with as much change as intertidal organisms do. Although the sublittoral zone extends down 200 m below sea level, plants and most animals stay in the upper 100 m, where small amounts of sunlight reaches the ocean floor.

The Bathyal Zone

The *bathyal* (BATH ee uhl) *zone* extends from the edge of the continental shelf to the abyssal plain. The depth of this zone ranges from 200 m to 4,000 m below sea level. Because of the lack of sunlight at these depths, plant life is scarce in this part of the benthic environment. Animals in this zone include sponges, brachiopods, sea stars, echinoids, and octopuses, such as the one shown in **Figure 4.**

Bathyal zone

Figure 4 *Octopuses are one of the animals common to the bathyal zone.*

Cultural Awareness — GENERAL

Sea Stories Stories of mermaids and sea monsters are a part of many cultures. In some legends, mermaids were good and helped shipwrecked sailors. In others, mermaids lured ships and sailors into dangerous waters. Sea monsters were always bad, sinking ships and killing sailors. Ask students why they think these legends were told. Have them research a legend and write one of their own. **LS Interpersonal**

SCIENCE HUMOR

Q: What's the best way to catch a fish?
A: Have someone throw it to you.

Figure 5 Tube worms can tolerate higher temperatures than most other organisms can. These animals survive in water as hot as 81°C.

Abyssal zone

The Abyssal Zone

No plants and very few animals live in the *abyssal zone*, which is on the abyssal plain. The abyssal zone is the largest ecological zone of the ocean and can reach 4,000 m in depth. Animals such as crabs, sponges, worms, and sea cucumbers live within the abyssal zone. Many of these organisms, such as the tube worms shown in **Figure 5,** live around hot-water vents called *black smokers.* Scientists know very little about this benthic environment because it is so deep and dark.

✓ Reading Check What types of animals live in the abyssal zone?

The Hadal Zone

The deepest benthic zone is the *hadal* (HAYD'l) *zone.* This zone consists of the floor of the ocean trenches and any organisms found there. The hadal zone can reach from 6,000 m to 7,000 m in depth. Scientists know even less about the hadal zone than they do about the abyssal zone. So far, scientists have discovered a type of sponge, a few species of worms, and a type of clam, which is shown in **Figure 6.**

Hadal zone

Figure 6 These clams are one of the few types of organisms known to live in the hadal zone.

Close

Reteaching — BASIC

Concept Mapping Have students create a concept map linking the two ocean environments and the ocean zones that make up each environment. **LS** Verbal

Quiz — GENERAL

1. Why do scientists know little about the abyssal and hadal zones? (Because these zones are so deep and dark, scientists know very little about them.)

2. Where do most marine organisms live? (in the neritic zone)

Alternative Assessment — GENERAL

Exploring the Abyssal Zone Challenge students to write science-fiction stories about exploring the abyssal zone of the benthic environment. Have them describe the difficulties of exploring that zone, and encourage them to use their imagination to describe creatures they might find there. **LS** Interpersonal

Neritic zone

Figure 7 *Many marine animals, such as these dolphins, live in the neritic zone.*

pelagic environment in the ocean, the zone near the surface or at middle depths, beyond the sublittoral zone and above the abyssal zone

The Pelagic Environment

The zone near the ocean's surface and at the middle depths of the ocean is called the **pelagic environment.** It is beyond the sublittoral zone and above the abyssal zone. There are two major zones in the pelagic environment—the neritic zone and the oceanic zone.

The Neritic Zone

The *neritic zone* covers the continental shelf. This warm, shallow zone contains the largest concentration of marine life. Fish, plankton, and marine mammals, such as the dolphins in **Figure 7,** are just a few of the animal groups found in this zone. The neritic zone contains diverse marine life because it receives more sunlight than the other zones in the ocean. Sunlight allows plankton, which are food for other marine organisms, to grow. The many animals in the benthic zone below the neritic zone also serve as a food supply.

✓ Reading Check Why does the neritic zone contain the largest concentration of marine life in the ocean?

CONNECTION TO Language Arts

WRITING SKILL **Water, Water, Everywhere** Samuel Taylor Coleridge wrote "The Rime of the Ancient Mariner" in 1798. The following is an excerpt from the poem:

Water, water, everywhere, / And all the boards did shrink / Water, water, everywhere, / Nor any drop to drink . . . / And every tongue through utter drought, / Was withered at the root; / We could not speak, no more than if / We had been choked with soot.

What do you think this excerpt means? Write a short essay describing the meaning of this passage.

Answer to Reading Check
The neritic zone contains the largest concentration of marine life in the ocean because it receives more sunlight than the other zones in the ocean.

Answer to Connection to Language Arts
Suggest that students use library resources or the Internet to help them research the meaning of the passage.

The Oceanic Zone

The *oceanic zone* includes the volume of water that covers the entire sea floor except for the continental shelf. In the deeper parts of the oceanic zone, the water temperature is colder and the pressure is much greater than in the neritic zone. Also, organisms are more spread out in the oceanic zone than in the neritic zone. Although many of the same organisms that live in the neritic zone are found throughout the upper regions, some strange animals lurk in the darker depths, as shown in **Figure 8.** Other animals in the deeper parts of this zone include giant squids and some whale species.

Oceanic zone

Figure 8 The angler fish is a predator that uses a wormlike lure attached to its head to attract prey.

SECTION Review

Summary

- The three main groups of marine life are plankton, nekton, and benthos.
- The two main ocean environments are the benthic environment and the pelagic environment.
- The ecological zones of the benthic environment include the intertidal zone, sublittoral zone, bathyal zone, abyssal zone, and hadal zone.
- The ecological zones of the pelagic environment include the neritic zone and the oceanic zone.

Using Key Terms

The statements below are false. For each statement, replace the underlined term to make a true statement.

1. <u>Plankton</u> are organisms that swim actively in ocean water.

2. The intertidal zone is part of the <u>pelagic zone.</u>

3. Dolphins live in the <u>benthic environment.</u>

Understanding Key Ideas

4. The deepest benthic zone is the
   a. pelagic environment.
   b. hadal zone.
   c. oceanic zone.
   d. abyssal zone.

5. List and briefly describe the three main groups of marine organisms.

6. Name the two ocean environments. In your own words, describe where they are located in the ocean.

Critical Thinking

7. **Making Inferences** Describe why organisms in the intertidal zone must be able to live underwater and on exposed land.

8. **Applying Concepts** How would the ocean's ecological zones change if sea level dropped 300 m?

Interpreting Graphics

Use the diagram below to answer the following question.

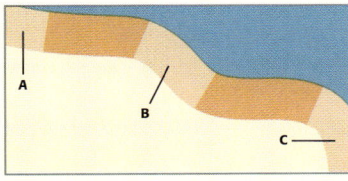

9. Identify the names of the ecological zones of the benthic environment shown above.

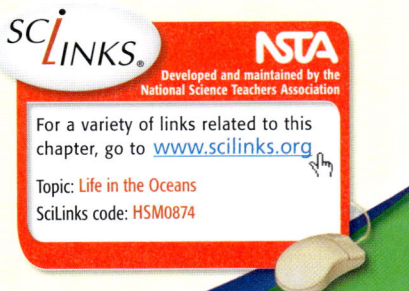

Developed and maintained by the National Science Teachers Association

For a variety of links related to this chapter, go to www.scilinks.org

Topic: Life in the Oceans
SciLinks code: HSM0874

Answers to Section Review

1. Nekton

2. benthic environment

3. pelagic environment

4. b

5. Sample answer: The three main groups of marine organisms are plankton, nekton, and benthos. Plankton are microscopic organisms that float or drift freely in freshwater and marine environments. Nekton are all organisms that swim actively in open water. Benthos are organisms that live at the bottom of the sea or ocean.

6. Sample answer: The two ocean environments are the benthic environment and the pelagic environment; The benthic environment is the bottom environment near the ocean floor and all the organisms that live on it or in it. The pelagic environment is the zone near the ocean's surface and at the middle depths of the ocean.

7. Sample answer: Organisms in the intertidal zone must be able to live underwater and exposed on land because twice a day, the intertidal zone changes. As the tide flows in, the zone is covered with ocean water, and as the tide flows out, the intertidal zone is exposed to the air and sun. Therefore, organisms in the intertidal zone must be able to adapt to these changes.

8. Sample answer: If sea level dropped 300 m, the shoreline would move down to the continental slope. This change would cause an overall shift of ecological zones because many marine organisms are restricted to certain depths.

9. A: intertidal zone, B: bathyal zone, C: hadal zone

Focus

Overview

This section discusses the ocean's living and nonliving resources and focuses on the methods used to obtain them. Students are asked to consider the importance of the ocean resources and to explore ways to conserve them.

🔔 Bellringer

Write the following sentences on the board, and challenge students to identify four items or activities that involve ocean resources.

• Tabitha drove her car to the market to buy a tuna steak for dinner. When she got home, she poured herself a glass of water, then fired up her gas grill, and cooked the tuna.

Motivate

Group ACTIVITY — GENERAL

Brainstorming Divide the class into small groups. Have students imagine a world without ocean resources. Ask them to brainstorm a list of activities that would no longer be possible. Examples might include eating seafood, shipping goods by sea, or deep-sea diving. Have each group share its list with the class and discuss the importance of ocean resources. **LS Verbal**

READING WARM-UP

Objectives
● List two ways of harvesting the ocean's living resources.
● Identify three nonliving resources in the ocean.
● Describe the ocean's energy resources.

Terms to Learn
desalination

READING STRATEGY

Paired Summarizing Read this section silently. In pairs, take turns summarizing the material. Stop to discuss ideas that seem confusing.

Resources from the Ocean

The next time you enjoy your favorite ice cream, remember that without seaweed, it would be a runny mess!

The ocean offers a vast supply of resources. These resources are put to a number of uses. For example, a seaweed called *kelp* is used as a thickener for many food products, including ice cream. Food, raw materials, energy, and drinkable water are all harvested from the ocean. And there are probably more resources in unexplored parts of the ocean. As human populations have grown, however, the demand for these resources has increased, while the availability has decreased.

Living Resources

People have been harvesting plants and animals from the ocean for thousands of years. Many civilizations formed in coastal regions where the ocean offered plenty of food for a growing population. Today, harvesting food from the ocean is a multi-billion-dollar industry.

Fishing the Ocean

Of all the marine organisms, fish are the largest group of organisms that are taken from the ocean. Almost 75 million tons of fish are harvested each year. With improved technology, such as drift nets, fishers have become better at taking fish from the ocean. **Figure 1** shows the large number of fish that can be caught using a drift net. In recent years, many people have become concerned that we are overfishing the ocean. We are taking more fish than can be naturally replaced. Also, animals other than fish, especially dolphins and turtles, can be accidentally caught in drift nets. Today, the fishing industry is making efforts to prevent overfishing and damage to other wildlife from drift nets.

Figure 1 *Drift nets are fishing nets that cover kilometers of ocean. Whole schools of fish can be caught with a single drift net.*

CHAPTER RESOURCES

Chapter Resource File

• **Lesson Plan**
• **Directed Reading A** BASIC
• **Directed Reading B** SPECIAL NEEDS

Technology

Transparencies
• Bellringer

Farming the Ocean

Overfishing reduces fish populations. Recently, laws regulating fishing have become stricter. As a result, it is becoming more difficult to supply our demand for fish. Many people have begun to raise ocean fish in fish farms to help meet the demand. Fish farming requires several holding ponds. Each pond contains fish at a certain level of development. **Figure 2** shows a holding pond in a fish farm. When the fish are old enough, they are harvested and packaged for shipping.

Fish are not the only seafood harvested in a farmlike setting. Shrimp, oysters, crabs, and mussels are raised in enclosed areas near the shore. Mussels and oysters are grown attached to ropes. Huge nets line the nursery area, preventing the animals from being eaten by their natural predators.

✔ **Reading Check** How can fish farms help reduce overfishing? (*See the Appendix for answers to Reading Checks.*)

 Figure 2 *Eating fish raised in a fish farm helps lower the number of fish harvested from the ocean.*

Savory Seaweed

Many types of seaweed, which are species of alga, are harvested from the ocean. For example, kelp, shown in **Figure 3,** is a seaweed that grows as much as 33 cm a day. Kelp is harvested and used as a thickener in jellies, ice cream, and similar products. Seaweed is rich in protein. In fact, several species of seaweed are staples of the Japanese diet. For example, some kinds of sushi, a Japanese dish, are wrapped in seaweed.

Figure 3 *Kelp, a type of alga, can grow up to 33 cm a day. It is harvested and used in a number of products, including ice cream.*

CONNECTION ACTIVITY
Environmental Science — ADVANCED

Factory Trawlers Factory trawlers are enormous boats that pull nets as large as four football fields and can harvest 400 tons of fish in a single haul. They can stay at sea for months catching, processing, freezing, and packaging fish. Because of their incredible cost (as much as $40 million), they must catch vast quantities of fish to remain operational. Factory trawlers threaten marine ecosystems because the trawlers can deplete an entire local fish population and quickly move on. They are also criticized for netting large quantities of *bycatch*, unwanted fish and sea life that are caught and thrown overboard, often dead or dying. However, fish are an important source of protein. Also, fishing is a major source of employment in the world, but many scientists agree that strong restrictions limiting fishing are important. Have students find out more about these ships and the environmental controversies that surround them. **LS** Verbal

Petroleum Dependence

Have students make a list of all their daily activities that rely on petroleum. (Answers might include taking a hot shower, using a hair dryer, cooking breakfast, driving to school, turning on lights, and washing and drying clothes.) Allow time for students to share their lists with classmates, and then encourage them to brainstorm ways they might reduce their reliance on fossil fuels.

Group ACTIVITY ——— GENERAL

Writing

Public Service Announcement

Divide the class into small groups, and encourage each group to select a different ocean resource. Have students work together to write a public service announcement designed to convince the public of the resource's value to people. Have them include ways people can help conserve the resource. Ask students to present their announcements to the class.

LS Verbal/Intrapersonal

Nonliving Resources

Humans also harvest many nonliving resources from the ocean. These resources provide raw materials, drinkable water, and energy for our growing population. Some resources are easy to get, while others are very difficult to harvest.

Oil and Natural Gas

Modern civilization continues to be very dependent on oil and natural gas for energy. Oil and natural gas are *nonrenewable resources*. They are used up faster than they can be replenished naturally. Both oil and natural gas are found under layers of impermeable rock. Petroleum engineers must drill through this rock in order to reach these resources.

✓ **Reading Check** What are nonrenewable resources? Give an example of a nonrenewable resource.

Searching for Oil

How do engineers know where to drill for oil and natural gas? They use seismic equipment. Special devices send powerful pulses of sound to the ocean floor. The pulses move through the water and penetrate the rocks and sediment below. The pulses are then reflected back toward the ship, where they are recorded by electronic equipment and analyzed by a computer. The computer readings indicate how rock layers are arranged below the ocean floor. Petroleum workers, such as the one in **Figure 4,** use these readings to locate a promising area to drill.

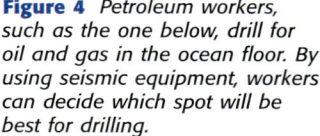

Figure 4 *Petroleum workers, such as the one below, drill for oil and gas in the ocean floor. By using seismic equipment, workers can decide which spot will be best for drilling.*

Homework ——— GENERAL

Graphing Ocean Resources Help students use references to find out how much of the world's energy needs are met by the following: oil, natural gas, coal, tidal energy, geothermal energy, wave energy, hydroelectric energy, solar energy, and wind energy. Have students prepare pie graphs or bar graphs of their findings. **LS** Visual/Interpersonal

Figure 5 Most desalination plants, like this one in Kuwait, use evaporation to separate ocean water from the salt it contains.

Fresh Water and Desalination

In parts of the world where fresh water is limited, people desalinate ocean water. **Desalination** (DEE SAL uh NAY shuhn) is the process of removing salt from sea water. After the salt is removed, the fresh water is then collected for human use. But desalination is not as simple as it sounds, and it is very expensive. Countries with enough annual rainfall rely on the fresh water provided by precipitation and do not need costly desalination plants. Some countries located in drier parts of the world must build desalination plants to provide enough fresh water. One of these plants is shown in **Figure 5.** Saudi Arabia, located in the desert region of the Middle East, has one of the largest desalination plants in the world.

desalination a process of removing salt from ocean water

Reading Check Explain where desalination plants are most likely to be built.

The Desalination Plant

1. Measure **1,000 mL of warm water** in a **graduated cylinder.** Pour the water in a **large pot.**

2. Carefully, add **35 g of table salt.** Stir the water until all of the salt is dissolved.

3. Place the pot on a **hot plate,** and allow all of the water to boil away.

4. Using a **wooden spoon,** scrape the salt residue from the bottom of the pot.

5. Measure the mass of the salt that was left in the bottom of the pot. How much salt did you separate from the water?

6. How does this activity model what happens in a desalination plant? What would be done differently in a desalination plant?

MATERIALS

FOR EACH GROUP
- graduated cylinder
- hot plate
- pot, large
- spoon, wooden
- table salt, 35 g
- water, warm, 1,000 mL

Safety Caution: Remind students to review all safety cautions and icons before beginning this lab activity.

Answers

5. Answers may vary.

6. Sample answer: In this activity, salt was separated from salt water by evaporation, just as in a desalination plant. In a desalination plant, the evaporating water would have been collected.

Quiz — GENERAL

1. Why are oil and natural gas considered nonrenewable resources? (because they can be used up faster than they can be replenished naturally)

2. Why does tidal energy generate a large amount of energy? (The motion of tides moves an enormous mass of water, which generates a large amount of energy.)

Alternative Assessment — GENERAL

 Depending on Ocean Resources Ask students to consider the ways in which their lives depend on ocean resources. Have students write a short essay about how their lives would be different if ocean resources did not exist. **LS Verbal/Interpersonal**

Figure 6 *Manganese nodules are difficult to mine because they are located on the deep ocean floor.*

Sea-Floor Minerals

Mining companies are interested in mineral nodules that are lying on the ocean floor. These nodules are made mostly of manganese, which can be used to make certain types of steel. They also contain iron, copper, nickel, and cobalt. Other nodules are made of phosphates, which are used to make fertilizer.

Nodules are formed from dissolved substances in sea water that stick to solid objects, such as pebbles. As more substances stick to the coated pebble, a nodule begins to grow. Manganese nodules can be as small as a marble or as large as a soccer ball. The photograph in **Figure 6** shows a number of nodules on the ocean floor. Scientists estimate that 15% of the ocean floor is covered with these nodules. However, these nodules are located in the deeper parts of the ocean, and mining them is costly and difficult.

Tidal Energy

The ocean generates a great deal of energy simply because of its constant movement. The gravitational pulls of the sun and moon cause the ocean to rise and fall as tides. *Tidal energy* is energy generated from the movement of tides. Tidal energy can be an excellent source of power. If the water during high tide can be rushed through a narrow coastal passageway, the water's force can be powerful enough to generate electrical energy. **Figure 7** shows how this process works. Tidal energy is a clean, inexpensive, and renewable resource. A *renewable resource* can be replenished, in time, after being used. Unfortunately, tidal energy is practical only in a few parts of the world. These areas must have a coastline with shallow, narrow channels. For example, the coastline at Cook Inlet, in Alaska, is ideal for generating electrical energy.

Figure 7 **Using Tides to Generate Electrical Energy**

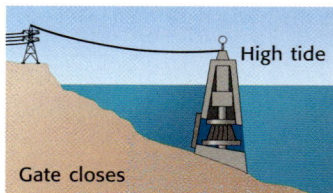

❶ As the tide rises, water enters a bay behind a dam. The gate then closes at high tide.

❷ The gate remains closed as the tide lowers.

❸ At low tide, the gate opens, and the water rushes through the dam and moves the turbines, which, in turn, generate electrical energy.

CONNECTION to Geology — GENERAL

Nodule Formation Explain that the mineral nodules shown in **Figure 6** form much as pearls or rock candy forms—a solid precipitates out of a chemical solution and adheres to a particle. This particle could be a small pebble or even a piece of shell. In a process called *accretion,* the nodules grow larger as more solids precipitate out. Students could model accretion by making rock candy with a supersaturated sugar solution and a piece of string. **LS Verbal**

Wave Energy

Have you ever stood on the beach and watched as waves crashed on the shore? This constant motion is an energy resource. Wave energy, like tidal energy, is a clean, renewable resource. Recently, computer programs have been developed to analyze wave energy. Researchers have found certain areas of the world where wave energy can generate enough electrical energy to make building power plants worthwhile. Wave energy in the North Sea is strong enough to produce power for parts of Scotland and England.

✓ Reading Check Why would wave energy be a good alternative energy resource?

SECTION Review

Summary

- Humans depend on the ocean for living and nonliving resources.
- Fish and other marine life are being raised in ocean farms to help feed growing human populations.
- Nonliving ocean resources include oil and natural gas, water, minerals, and tidal and wave energy.

Using Key Terms

1. In your own words, write a definition for the term *desalination*.

Understanding Key Ideas

2. Mineral nodules on the ocean floor are
   a. renewable resources.
   b. easily mined.
   c. used during the process of desalination.
   d. nonliving resources.

3. List two ways of harvesting the ocean's living resources.

4. Name four nonliving resources in the ocean.

5. Explain how fish farms help meet the demand for fish.

6. Explain how engineers decide where to drill for oil and natural gas in the ocean.

Math Skills

7. A kelp plant is 5 cm tall. If it grows an average of 29 cm per day, how tall will the kelp plant be after 2 weeks?

Critical Thinking

8. **Analyzing Processes** Explain why tidal energy and wave energy are considered renewable resources.

9. **Predicting Consequences** Define the term *overfishing* in your own words. What would happen to the population of fish in the ocean if laws did not regulate overfishing? What would happen to the ocean ecosystem?

10. **Analyzing Ideas** What is one benefit and one consequence of building a desalination plant? Would a desalination plant be beneficial to your local area? Explain why or why not.

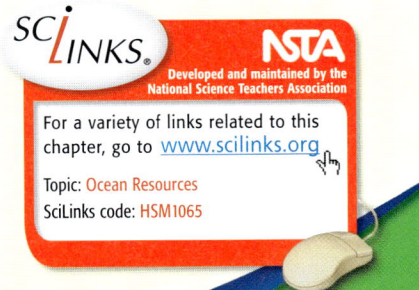

For a variety of links related to this chapter, go to www.scilinks.org

Topic: Ocean Resources
SciLinks code: HSM1065

Answers to Section Review

1. Sample answer: Desalination is a process for removing salt from ocean water.

2. d

3. Sample answer: Two ways of harvesting the ocean's living resources include fishing the ocean for marine organisms and farming the ocean by raising marine organisms in fish or seaweed farms.

4. oil, fresh water, sea-floor minerals, and energy

CHAPTER RESOURCES

Chapter Resource File

- Section Quiz **GENERAL**
- Section Review **GENERAL**
- Vocabulary and Section Summary **GENERAL**
- Reinforcement Worksheet **BASIC**
- Critical Thinking **ADVANCED**
- Datasheet for Quick Lab

---

5. Sample answer: Fish farms help meet the demand for fish by breeding fish in farms instead of fishing them directly from the ocean.

6. Sample answer: Engineers decide where to drill for oil and natural gas in the ocean by using seismic equipment. This equipment sends powerful pulses of sound from a ship to the ocean floor and then back to the ship. These pulses are then analyzed by a computer. The pulses indicate how rock layers are arranged below the ocean floor and therefore where there are good locations for drilling.

7. (14 days × 29 cm) = 406 cm, (406 cm + 5 cm) = 411 cm

8. Sample answer: Tidal energy and wave energy are renewable resources because tides and waves are naturally occurring movements of water. Therefore, as long as tides and waves continue to occur on Earth, energy can be harnessed from the movement of tides and waves.

9. Sample answer: Overfishing occurs when we take more fish from the ocean than can be naturally replaced; If laws did not regulate overfishing, the fish population would continue to decrease; If overfishing continued, ocean food chains, food webs, and ocean ecosystems would eventually collapse.

10. Answers may vary. Sample answer: One benefit of building a desalination plant is that it allows an area that is very dry to have a supply of fresh water. One consequence is that it is very expensive.

Answer to Reading Check

Wave energy would be a good alternative energy resource because it is a clean and renewable resource.

Focus

Overview

In this section, students learn about the sources and effects of ocean pollution. They also learn about some of the strategies that are used to minimize ocean pollution.

 Bellringer

Ask students to write a few sentences about how ocean pollution could affect their lives. Then, have students list the ways they contribute to ocean pollution in their daily lives. Ask groups to brainstorm about the ways they could reduce ocean pollution.

Motivate

ACTiViTY ———————— GENERAL

Cleaning Up an Oil Spill

Give each pair of students a pan with water. Pour about 5 mL of vegetable oil into each pan. Have students think of ways to remove the oil from the pan *without pouring out the water*. Students can experiment with methods such as absorbing the oil with a paper towel or scooping it out with a spoon. Discuss the results of students' efforts. Tell them that this problem is similar to dealing with a petroleum spill in the ocean. **LS** Visual

READING WARM-UP

Objectives

● Explain the difference between point-source pollution and nonpoint-source pollution.

● Identify three different types of point-source ocean pollution.

● Describe what is being done to control ocean pollution.

Terms to Learn

nonpoint-source pollution
point-source pollution

READING STRATEGY

Reading Organizer As you read this section, create an outline of the section. Use the headings from the section in your outline.

Ocean Pollution

It's a hot summer day at the beach. You can hardly wait to swim in the ocean. You run to the surf only to be met by piles of trash washed up on the shore. Where did all that trash come from?

Humans have thrown their trash in the ocean for hundreds, if not thousands, of years. This trash has harmed the plants and animals that live in the oceans, as well as the people and animals that depend on them. Fortunately, we are becoming more aware of ocean pollution, and we are learning from our mistakes.

Nonpoint-Source Pollution

There are many sources of ocean pollution. Some of these sources are easily identified, but others are more difficult to pinpoint. **Nonpoint-source pollution** is pollution that comes from many sources rather than just from a single site. Some common sources of nonpoint-source pollutants are shown in **Figure 1.** Most ocean pollution is nonpoint-source pollution. Human activities on land can pollute streams and rivers, which then flow into the ocean and bring the pollutants they carry with them. Because nonpoint-source pollutants can enter bodies of water in many different ways, they are very hard to regulate and control. Nonpoint-source pollution can be reduced by using less lawn chemicals and disposing of used motor oil properly.

Figure 1 Examples of Nonpoint-Source Pollution

Oil and gasoline that have leaked from cars onto streets can wash into storm sewers and then drain into waterways.

Thousands of watercraft, such as boats and jet skis can leak gasoline and oil directly into bodies of water.

Pesticides, herbicides, and fertilizer from residential lawns, golf courses, and farmland can wash into waterways.

CHAPTER RESOURCES

Chapter Resource File

 • Lesson Plan
• Directed Reading A BASIC
• Directed Reading B SPECIAL NEEDS

Technology

 Transparencies
• Bellringer

SCIENCE HUMOR

Huntsville, Alabama, hairdresser Phil McCrory has patented a way to use discarded human hair to clean up oil spills. He tested the idea in his son's wading pool by using a pair of pantyhose filled with hair. The idea works so well that McCrory has attracted the attention of NASA!

Figure 2 *This barge is headed out to the open ocean, where it will dump the trash it carries.*

Point-Source Pollution

Water pollution caused by a leaking oil tanker, a factory, or a wastewater treatment plant is one type of point-source pollution. **Point-source pollution** is pollution that comes from a specific site. Even when the source of pollution is known, cleanup of the pollution is difficult.

Trash Dumping

People dump trash in many places, including the ocean. In the 1980s, scientists became alarmed by the kinds of trash that were washing up on beaches. Bandages, vials of blood, and syringes (needles) were found among the waste. Some of the blood in the vials even contained the AIDS virus. The Environmental Protection Agency (EPA) began an investigation and discovered that hospitals in the United States produce an average of 3 million tons of medical waste each year. Because of stricter laws, much of this medical waste is now buried in sanitary landfills. However, dumping trash in the deeper part of the ocean is still a common practice in many countries. The barge in **Figure 2** will dump the trash it carries into the open ocean.

Effects of Trash Dumping

Trash thrown into the ocean can affect the organisms that live in the ocean and those organisms that depend on the ocean for food. Trash such as plastic can be particularly harmful to ocean organisms. This is because most plastic materials do not break down for thousands of years. Marine animals can mistake plastic materials for food and choke or become strangled. The sea gull in **Figure 3** is tangled up in a piece of plastic trash.

☑️ **Reading Check** What is one effect of trash dumping? (*See the Appendix for answers to Reading Checks.*)

nonpoint-source pollution pollution that comes from many sources rather than from a single, specific site

point-source pollution pollution that comes from a specific site

Figure 3 *Marine animals can be strangled by plastic trash or can choke if they mistake the plastic for food.*

Answer to Reading Check

One effect of trash dumping is that plastic materials may harm and kill marine animals because these animals may mistake the trash for food.

Is That a Fact!

The world's first major oil spill occurred on March 18, 1967. The tanker *Torrey Canyon* ran aground off the coast near Cornwall, England. The ship spilled about 870,000 barrels of oil, more than 3 times the oil spilled by the *Exxon Valdez*.

Teach

ACTIVITY ————— BASIC

Nonpoint-Source Pollution
Draw students' attention to the photographs of nonpoint-source pollutants in **Figure 1.** Ask them to think about their daily lives and to list ways they might be contributing to nonpoint-source pollution. Ask them to help you list ways that they could reduce their contribution to nonpoint-source pollution. **LS Verbal** ⬛ **English Language Learners**

Homework ————— GENERAL

Investigate Your Area Inform students that many companies are creating products that are marketed to be less harmful to the environment. These claims have varying degrees of truth. Have groups choose a product and investigate the environmental statements that are used to market it. For example, a group could visit the detergent aisle of a grocery store to compare products. Have students prepare a list of products claiming to be environmentally friendly and their primary ingredients. Encourage students to use Internet or library resources to determine whether the ingredients listed could be harmful to the environment. Another group could compare the percentage of postconsumer waste in recycled paper products. Have groups present their findings to the class. **LS Intrapersonal**

CONNECTION ACTIVITY
Math — GENERAL

Global Ocean Pollution Challenge students to use the Internet or library resources to find out which nations contribute the most to ocean pollution. Have students construct pie graphs to show their results and write a short essay explaining the trends they observe. **LS Visual**

Discussion — GENERAL

Nuclear Waste and the Ocean Remind students that the ocean tends to separate into layers because temperature and salinity differences create layers with different densities.

Point out that the world community is confronted with the dilemma of nuclear waste disposal. One proposed solution is to bury weighted barrels of waste in the sediment of the ocean floor. Encourage students to consider the implications of such a strategy. As nuclear waste decays, it releases heat energy. What would happen if water contaminated by nuclear waste heated the deep zone? **LS Verbal**

Figure 4 *Sludge is the solid part of waste matter and often carries bacteria. Sludge makes beaches dirty and kills marine animals.*

Sludge Dumping

By 1990, the United States alone had discharged 38 trillion liters of treated sludge into the waters along its coasts. Sludge is part of raw sewage. *Raw sewage* is all the liquid and solid wastes that are flushed down toilets and poured down drains. After collecting in sewer drains, raw sewage is sent through a treatment plant, where it undergoes a cleaning process that removes solid waste. The solid waste is called *sludge,* as shown in **Figure 4.** In many areas, people dump sludge into the ocean several kilometers offshore, intending for it to settle and stay on the ocean floor. Unfortunately, currents can stir the sludge up and move it closer to shore. This sludge can pollute beaches and kill marine life. Many countries have banned sludge dumping, but it continues to occur in many areas of the world.

Oil Spills

Because oil is in such high demand across the world, large tankers must transport billions of barrels of it across the oceans. If not handled properly, these transports can turn disastrous and cause oil spills. **Figure 5** shows some of the major oil spills that have occurred off the coast of North America.

Figure 5 *This map shows some of the major oil spills that have occurred off the coast of North America in the last 30 years.*

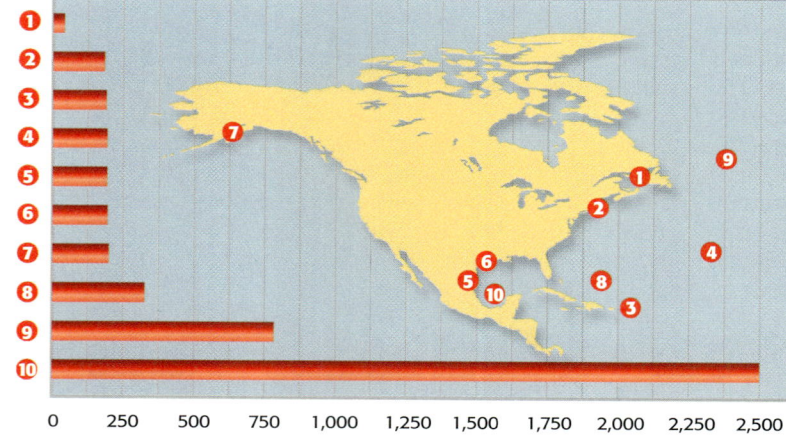

Barrels spilled (in thousands)

❶ *Kurdistan* Gulf of St. Lawrence, Canada, 1979

❷ *Argo Merchant* Nantucket, MA, 1976

❸ Storage Tank Benuelan, Puerto Rico, 1978

❹ *Athenian Venture* Atlantic Ocean, 1988

❺ Unnamed Tanker Tuxpan, Mexico, 1996

❻ *Burmah Agate* Galveston Bay, TX, 1979

❼ *Exxon Valdez* Prince William Sound, AK, 1989

❽ *Epic Colocotronis* Caribbean Sea, 1975

❾ *Odyessey* North Atlantic Ocian, 1988

❿ Exploratory Well Bay of Campeche, 1979

Homework — GENERAL

Environmental Impacts Help students find information to investigate the environmental impact of one of the following: developing areas where salt marshes once were, draining mangrove swamps for development, polluting of ocean waters with agricultural runoff, damming rivers and streams that flow into the ocean, using dynamite to catch fish (common in the Pacific and particularly harmful to coral reefs), or destroying coral reefs to obtain construction materials. Encourage students to focus on the effects of these human behaviors on other organisms, and challenge them to suggest alternatives **LS Verbal**

Effects of Oil Spills

One of the oil spills shown on the map in **Figure 5** occurred in Prince William Sound, Alaska, in 1989. The supertanker *Exxon Valdez* struck a reef and spilled more than 260,000 barrels of crude oil along the shorelines of Alaska. The amount of spilled oil is roughly equivalent to 125 olympic-sized swimming pools.

Although some animals were saved, such as the bird in **Figure 6,** many plants and animals died as a result of the spill. Alaskans who made their living from fishing lost their businesses. The Exxon Oil Company spent $2.1 billion to try to clean up the mess. But Alaska's wildlife and economy will continue to suffer for decades.

While oil spills can harm plants, animals, and people, they are responsible for only about 5% of oil pollution in the oceans. Most of the oil that pollutes the oceans is caused by nonpoint-source pollution on land from cities and towns.

Figure 6 *Many oil-covered animals were rescued and cleaned after the* Exxon Valdez *spill.*

Preventing Oil Spills

Today, many oil companies are using new technology to safeguard against oil spills. Tankers are now being built with two hulls instead of one. The inner hull prevents oil from spilling into the ocean if the outer hull of the ship is damaged. **Figure 7** shows the design of a double-hulled tanker.

✓ Reading Check How can two hulls on an oil tanker help prevent an oil spill?

Figure 7 *If the outer hull of a double-hulled tanker is punctured, the oil will still be contained within the inner hull.*

CONNECTION to Environmental Science — GENERAL

Double-Hulled Tankers The Oil Pollution Act of 1990 was a direct response to the *Exxon Valdez* oil spill. The controversial bill had been debated for 14 years; it passed swiftly in the aftermath of the disaster. Under the law, all oil tankers operating in United States waters must be double-hulled by 2015. Compliance has been slow, however; many oil companies have been reluctant to replace the aging boats in their fleets with hundred-million-dollar double-hulled ships. As of 1999, of the 3,294 oil tankers operating worldwide, only 876 were double hulled.

ACTIVITY — GENERAL

Ocean Pollution Awareness
Divide the class into small groups. Provide each group with poster board and markers. Direct each group to work together to create a poster designed to educate the public about the need for clean water. Ask groups to focus on ways people can minimize the pollution of our oceans. Consider displaying the posters around the school to educate students and teachers. As an extension, have students draft a letter to a United States Congress member that outlines each group's ideas. **English Language Learners**
LS Intrapersonal

Group ACTIVITY — GENERAL

Exxon Valdez Have student groups report on the current status of the communities and ecosystems of Prince William Sound in the wake of the *Exxon Valdez* oil spill. Ask students to consider questions such as the following: "How has the ecosystem recovered? Do fishing communities feel that they have been compensated?"
LS Intrapersonal

Answer to Reading Check
An oil tanker that has two hulls can prevent an oil spill, because if the outer hull is damaged, the inner hull will prevent oil from spilling into the ocean.

Reteaching ——— BASIC

Effects of Ocean Pollution
Have students list one type of pollution they saw on the way to school today. Ask them if this type of pollution affects the ocean. **LS** Verbal

Quiz ——— GENERAL

1. Why do oil spills pose a long-term risk? (Sample answer: Organisms living in the ocean are part of a complex web of relationships. If one group of organisms is affected, many other groups could be threatened. In addition, the oil tends to concentrate in the tissues of living organisms and can persist long after the spill has been cleaned up.)

2. Why is sludge dumping a threat to marine and human life? (Sludge dumping can cause diseases, kill marine organisms, and pollute beaches.)

Alternative Assessment ——— GENERAL

Concept Mapping Have students recall the sources of ocean pollution discussed in this section, the effects of pollution on oceans and wildlife, and the actions being taken by nations and individuals to limit and reduce pollution. Ask them to prepare a concept map that organizes and compares this information. Challenge students to propose additional suggestions for pollution prevention.
LS Visual/Verbal

SCHOOL to HOME

Coastal Cleanup
WRITING SKILL You can be a part of a coastal cleanup. Every September, people from all over the world set aside one day to help clean up trash and debris from beaches. You can join this international effort! With a parent, write a letter to the Ocean Conservancy to see what you can do to help clean up coastal areas.

ACTIVITY

Figure 8 *Making an effort to pick up trash on a beach can help make the beach safer for plants, animals, and people.*

Saving Our Ocean Resources

Although humans have done much to harm the ocean's resources, we have also begun to do more to save them. From international treaties to volunteer cleanups, efforts to conserve the ocean's resources are making an impact around the world.

Nations Take Notice

When ocean pollution reached an all-time high, many countries recognized the need to work together to solve the problem. In 1989, a treaty was passed by 64 countries that prohibits the dumping of certain metals, plastics, oil, and radioactive wastes into the ocean. Even though many other international agreements and laws restricting ocean pollution have been made, waste dumping and oil spills still occur. Therefore, waste continues to wash ashore, as shown in **Figure 8.** Enforcing pollution-preventing laws at all times is often difficult.

Citizens Taking Charge

Citizens of many countries have demanded that their governments do more to solve the growing problem of ocean pollution. Because of public outcry, the United States now spends more than $130 million each year to protect the oceans and beaches. United States citizens have also begun to take the matter into their own hands. In the early 1980s, citizens began organizing beach cleanups. One of the largest cleanups is the semiannual Adopt-a-Beach program, shown in **Figure 8,** which originated with the Texas Coastal Cleanup campaign. Millions of tons of trash have been gathered from the beaches, and people are being educated about the hazards of ocean dumping.

Group ACTIVITY — GENERAL

Coastal Campaign Divide the class into small groups. Ask groups to imagine that they are the campaign managers for a candidate trying to become the city manager of a coastal town. Tell them that they must convince voters that the groups' candidates are concerned for the environmental health of the ocean and have a realistic plan to preserve the beaches and marine life. Have them write a campaign speech outlining the sources of pollution and the solutions they propose. Ask one student from each group to present the speech to the class. Allow the class to vote for the most convincing candidate.
LS Verbal

Action in the United States

The United States, like many other countries, has taken additional measures to control local pollution. For example, in 1972, Congress passed the Clean Water Act, which put the Environmental Protection Agency in charge of issuing permits for any dumping of trash into the ocean. Later that year, a stricter law—the U.S. Marine Protection, Research, and Sanctuaries Act—was passed. This act prohibits the dumping of any material that would affect human health or welfare, the marine environment or ecosystems, or businesses that depend on the ocean.

Reading Check What is the U.S. Marine Protection, Research, and Sanctuaries Act?

SECTION Review

Summary

- The two main types of ocean pollution are nonpoint-source pollution and point-source pollution.
- Types of nonpoint-source pollution include oil and gasoline from cars, trucks, and watercraft, as well as the use of pesticides, herbicides, and fertilizers.
- Types of point-source ocean pollution include trash dumping, sludge dumping, and oil spills.
- Efforts to save ocean resources include international treaties and volunteer cleanups.

Using Key Terms

1. Use the following terms in the same sentence: *point-source pollution* and *nonpoint-source pollution*.

Understanding Key Ideas

2. Which of the following is an example of nonpoint-source pollution?
   a. a leak from an oil tanker
   b. a jet ski
   c. an unlined landfill
   d. water discharged by industries

3. List three types of ocean pollution. How can each of these types be prevented or minimized?

4. Which part of raw sewage is a type of ocean pollution?

Math Skills

5. Only 3% of Earth's water is drinkable. What portion of Earth's water is not drinkable?

6. A ship spilled 750,000 barrels of oil when it accidentally struck a reef. The oil company was able to recover 65% of the oil spilled. How many barrels of oil were not recovered?

Critical Thinking

7. **Identifying Relationships** List and describe three measures that governments have taken to control ocean pollution.

8. **Evaluating Data** What were two effects of the *Exxon Valdez* oil spill? Describe two ways in which oil spills can be prevented.

9. **Applying Concepts** List two examples of nonpoint-source pollution that occur in your area. Explain why they are nonpoint-source pollution.

10. **Predicting Consequences** How can trash dumping and sludge dumping affect food chains in the ocean?

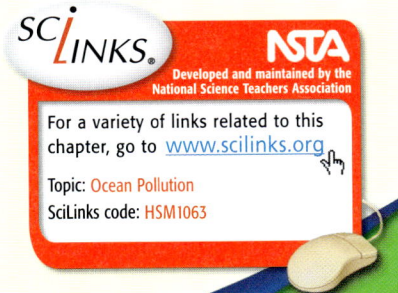

SCI LINKS®

Developed and maintained by the National Science Teachers Association

For a variety of links related to this chapter, go to www.scilinks.org

Topic: Ocean Pollution
SciLinks code: HSM1063

Answers to Section Review

1. Sample answer: Trash dumping is an example of point-source pollution, and the use of pesticides is an example of nonpoint-source pollution.

2. d

3. Sample answer: trash dumping, sludge dumping, and oil spills; Trash can be dumped in sanitary landfills, sludge can be used to make compost, and tankers with double hulls can be used to transport oil. All of these types of pollution can be prevented or minimized through legislation, education, and actions by citizens.

CHAPTER RESOURCES

Chapter Resource File

- Section Quiz **GENERAL**
- Section Review **GENERAL**
- Vocabulary and Section Summary **GENERAL**

4. sludge

5. 97%

6. (750,000 barrels × 0.65) = 487,500 barrels, (750,000 barrels − 487,500 barrels) = 262,500 barrels were not recovered

7. Three measures taken by governments to control ocean pollution include passing laws that prohibit the dumping of radioactive wastes into the ocean, that prohibit the dumping of trash into the ocean, and that prohibits dumping of any material that would affect any organism that depends on the ocean.

8. Two effects of the *Exxon Valdez* oil spill were that many plants and animals died and that many Alaskans, who made their living from fishing, lost their businesses. Two ways in which oil spills can be prevented are to reduce shipping oil by sea and to build oil tankers with two hulls.

9. Sample answer: using watercraft such as boats and jet skis and using pesticides on lawns; The use of watercraft can be minimized, and humans can use cleaner fuels to operate their watercraft. The use of pesticides on lawns can be reduced, and humans can use more environmentally friendly ways of pest control.

10. Trash dumping and sludge dumping can affect a food chain because both trash and toxic materials from sludge can make organisms sick or die. Therefore, if organisms eat other organisms that are sick, they will likely become sick, too. Also, if organisms die, animals that depend on those animals for food may starve.

Answer to Reading Check

The U.S. Marine Protection, Research, and Sanctuaries Act prohibits the dumping of any material that would affect human health or welfare, the marine environment or ecosystems, or businesses that depend on the ocean.

Probing the Depths

Teacher's Notes

Time Required
One 45-minute class period

Lab Ratings
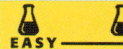

EASY ——————→ HARD

Teacher Prep 🧪
Student Set-Up 🧪🧪
Concept Level 🧪🧪
Clean Up 🧪

MATERIALS
The materials listed on the student page are enough for each student or for a group of 2 to 4 students.

Safety Caution
Remind students to review all safety cautions and icons before beginning this lab activity.

Preparation Notes
You may wish to ask students to provide their own shoe boxes. Punch the holes along the center of the lids for the students prior to class. You may use corrugated cardboard instead of modeling clay. If you do, shape the cardboard into steps and ridges to model changes in depth. With this type of model, students should use the eraser end of an unsharpened pencil to measure depths.

Model-Making Lab

Model-Making Lab

OBJECTIVES

Model a method of mapping the ocean floor.

Construct a map of an ocean-floor model.

MATERIALS

- clay, modeling (1 lb)
- pencil, unsharpened (8 of equal length)
- ruler, metric
- scissors
- shoe box with lid

SAFETY

Probing the Depths

In the 1870s, the crew of the ship the HMS *Challenger* used a wire and a weight to discover and map some of the deepest places in the world's oceans. The crew members tied a wire to a weight and dropped the weight overboard. When the weight reached the bottom of the ocean, they hauled the weight back up to the surface and measured the length of the wet wire. In this way, they were eventually able to map the ocean floor. In this activity, you will model this method of mapping by making a map of an ocean-floor model.

Procedure

1 Use the clay to make a model ocean floor in the shoe box. The model ocean floor should have some mountains and valleys.

2 Cut eight holes in a line along the center of the lid. The holes should be just big enough for a pencil to slide through. Place the lid on the box.

3 Exchange boxes with another student or group of students. Do not look into the box.

4 Copy the table shown on the facing page onto a piece of paper. Also, copy the graph shown on the facing page.

5 Measure the length of the probe (pencil) in centimeters. Record the length in your data table.

6 Gently insert the probe into the first hole position in the box until the probe touches the model ocean floor. Do not push the probe down. Pushing the probe down could affect your reading.

7 Make sure that the probe is straight up and down, and measure the length of probe showing above the lid. Record your data in the data table.

8 Use the formula below to calculate the depth in centimeters.

$$\text{length of probe} - \text{length of probe showing (cm)} = \text{depth (cm)}$$

CHAPTER RESOURCES

Chapter Resource File
 • Datasheet for Chapter Lab
• Lab Notes and Answers

Technology
 Classroom Videos
• Lab Video

 LabBook

• Investigating an Oil Spill

Ocean Depth Table				
Hole position	Length of probe	Length of probe showing (cm)	Depth (cm)	Depth (m) scale of 1cm = 200m
1				
2				
3				
4				
5		DO NOT WRITE IN BOOK		
6				
7				
8				

9 To better represent real ocean depths, use the scale 1 cm = 200 m to convert the depth in centimeters to depth in meters. Add the data to your table.

10 Plot the depth in meters for hole position 1 on your graph.

11 Repeat steps 6–10 for the other hole positions.

12 After plotting the data for the eight hole positions, connect the plotted points with a smooth curve.

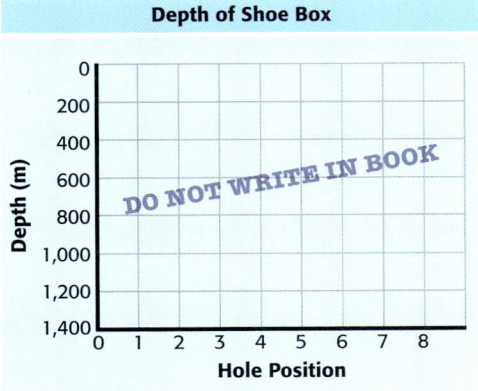

Depth of Shoe Box

DO NOT WRITE IN BOOK

13 Put a pencil in each of the holes in the shoe box. Compare the rise and fall of the eight pencils with the shape of your graph.

Analyze the Results

1 **Describing Events** How deep was the deepest point of your ocean-floor model? How deep was the shallowest point of your ocean-floor model?

2 **Explaining Events** Did your graph resemble the ocean-floor model, as shown by the pencils in step 13? If not, why not?

Draw Conclusions

3 **Applying Conclusions** Why is measuring the real ocean floor difficult? Explain your answer.

CHAPTER RESOURCES

Workbooks

Whiz-Bang Demonstrations
• Foul Play GENERAL

Inquiry Labs
• Surf's Up! GENERAL

EcoLabs & Field Activities
• Rescue Near the Center of the Earth GENERAL

Long-Term Projects & Research Ideas
• Your Very Own Underwater Theme Park ADVANCED

Calculator-Based Labs
• Ocean Floor Mapping ADVANCED

Chapter Review

Assignment Guide

SECTION	QUESTIONS
1	3, 7–10, 14, 19
2	1, 4, 15–16, 21
3	5, 12–13, 17, 23–26
4	2, 11, 18, 20
5	6, 22

ANSWERS

Using Key Terms

1. continental shelf
2. Desalination
3. Salinity
4. abyssal plain
5. benthic environment
6. nonpoint-source pollution

Understanding Key Ideas

7. b
8. c
9. d
10. c
11. c
12. a
13. a

USING KEY TERMS

Complete each of the following sentences by choosing the correct term from the word bank.

continental shelf
abyssal plain
salinity
nonpoint-source pollution
continental slope
desalination
benthic environment
point-source pollution

1. The region of the ocean floor that is closest to the shoreline is the ___.

2. ___ is the process of removing salt from sea water.

3. ___ is a measure of the amount of dissolved salts in a liquid.

4. The ___ is the broad, flat part of the deep-ocean basin.

5. The region near the bottom of a pond, lake, or ocean is called the ___.

6. Pollution that comes from many sources rather than a single specific source is called ___.

UNDERSTANDING KEY IDEAS

Multiple Choice

7. The largest ocean is the
   a. Indian Ocean.
   b. Pacific Ocean.
   c. Atlantic Ocean.
   d. Arctic Ocean.

8. One of the most abundant elements in the ocean is
   a. potassium.
   b. calcium.
   c. chlorine.
   d. magnesium.

9. Which of the following affects the ocean's salinity?
   a. fresh water added by rivers
   b. currents
   c. evaporation
   d. All of the above

10. Most precipitation falls
    a. on land.
    b. into lakes and rivers.
    c. into the ocean.
    d. in rain forests.

11. Which of the following is a non-renewable resource in the ocean?
    a. fish
    b. tidal energy
    c. oil
    d. All of the above

12. Which benthic zone has a depth range between 200 m and 4,000 m?
    a. the bathyal zone
    b. the abyssal zone
    c. the hadal zone
    d. the sublittoral zone

13. The ocean floor and all of the organisms that live on or in it is the
    a. benthic environment.
    b. pelagic environment.
    c. neritic zone.
    d. oceanic zone.

14. Coastal water in hotter, drier climates typically has higher salinity because less fresh water runs into the ocean in drier areas and because heat increases the evaporation rate.

15. Answers may vary. Sample answer: Technology used to study the ocean floor includes sonar and piloted vessels. *Sonar* stands for "sound navigation and ranging" and is used to determine the ocean's depth by sending sound pulses from a ship down into the ocean. Piloted vessels such as *Alvin* and *Deep Flight* are research vessels used to explore the ocean floor.

16. Sample answer: The two major regions of the ocean floor include the continental margin and the deep-ocean basin. The continental shelf, the continental slope, and the continental rise are sections of the continental margin based on depth and changes in slope.

Short Answer

14 Why does coastal water in areas that have hotter, drier climates typically have a higher salinity than coastal water in cooler, more humid areas does?

15 Describe two technologies used for studying the ocean floor.

16 Identify the two major regions of the ocean floor, and describe how the continental shelf, the continental slope, and the continental rise are related.

17 In your own words, write a definition for each of the following terms: *plankton*, *nekton*, and *benthos*. Give two examples of each.

18 List two living resources and two non-living resources that are harvested from the ocean.

CRITICAL THINKING

19 Concept Mapping Use the following terms to create a concept map: *water cycle, evaporation, condensation, precipitation, atmosphere,* and *oceans*.

20 Making Inferences What benefit other than being able to obtain fresh water from salt water comes from desalination?

21 Making Comparisons Explain the difference between a bathymetric profile and a seismic reading.

22 Analyzing Ideas In your own words, define *nonpoint-source pollution* and *point-source pollution*. Give an example of each. What is being done to control ocean pollution?

INTERPRETING GRAPHICS

The graph below shows the ecological zones of the ocean. Use the graph below to answer the questions that follow.

Ecological Zones of the Ocean

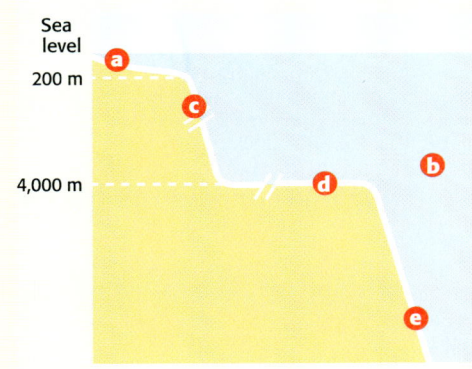

23 At which point would you most likely find an anglerfish?

24 At which point would you most likely find tube worms?

25 Which ecological zone is shown at point c? Which depth zone is shown at point c?

26 Name an organism that you might find at point e.

Critical Thinking

19. 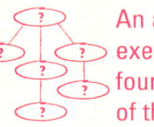 An answer to this exercise can be found at the end of this book.

20. Answers may vary. Sample answer: Desalination may also provide a source of minerals, such as salt.

21. Answers may vary. Sample answer: A bathymetric profile shows what the contour of the ocean floor looks like. A seismic reading can show what the contour of the ocean floor and what rocks beneath the ocean floor look like.

22. Sample answer: Nonpoint-source pollution is pollution that comes from many sources rather than from a single, specific site. Point-source pollution is pollution that comes from a specific site; An example of nonpoint-source pollution includes oil and gasoline that have leaked from cars onto streets. An example of point-source pollution is trash dumping from barges; Controlling ocean pollution is difficult, but many countries have taken legal action to protect oceans. For example, the United States passed the U.S. Marine Protection, Research, and Sanctuaries Act, which prohibits harmful dumping in the ocean. Beach cleanups run by citizens, such as the Adopt-a-Beach program, have been very successful in cleaning up polluted beaches.

Interpreting Graphics

23. b

24. d

25. bathyal zone; continental slope from 200 m to 4,000 m below sea level

26. Sample answer: clams

17. Sample answer: Plankton are microscopic organisms that float freely in freshwater and marine environments. Examples of plankton are phytoplankton and zooplankton. Nekton are organisms that swim actively in open water and include sea lions and whales. Benthos are the organisms that live at the bottom of the ocean, such as crabs and starfish.

18. Sample answer: Two living resources that are harvested from the ocean include fish and seaweed. Two nonliving resources that are harvested from the ocean include oil and sea-floor minerals.

Standardized Test Preparation

To provide practice under more realistic testing conditions, give students 20 minutes to answer all of the questions in this Standardized Test Preparation.

MISCONCEPTION ALERT

Answers to the standardized test preparation can help you identify student misconceptions and misunderstandings.

READING

Passage 1

1. C
2. I

TEST DOCTOR

Question 1: Answer A is incorrect because this problem is not mentioned in the passage. Answer B is incorrect because the passage implies that the cleanup was less than successful. Answer D is incorrect because Alaska's economy will suffer because of damage caused by the spill, not because of lost oil.

READING

Read each of the passages below. Then, answer the questions that follow each passage.

Passage 1 Because oil is in such high demand across the world, large tankers must transport billions of barrels of it across the oceans. If not handled properly, these transports can quickly turn disastrous. In 1989, the supertanker *Exxon Valdez* struck a reef and spilled more than 260,000 barrels of crude oil. The effect of this accident on wildlife was catastrophic. Within the first few weeks of the *Exxon Valdez* oil spill, more than half a million birds, including 109 endangered bald eagles, were covered with oil and drowned. Almost half the sea otters in the area also died, either from drowning or from being poisoned by the oil. Alaskans who made their living from fishing lost their businesses. Although many animals were saved and the Exxon Oil Company spent $2.1 billion to clean up the mess, Alaska's wildlife and economy will continue to suffer for decades.

1. What is the main idea of this passage?
   - **A** Transporting oil over long distances is difficult.
   - **B** The Exxon Oil Company did a great job of cleaning up the oil spill in Alaska.
   - **C** Oil spills such as the *Exxon Valdez* spill can create huge problems.
   - **D** Alaska's economy will suffer because so much oil was lost.

2. In the passage, which of the following problems was said to be a result of the *Exxon Valdez* oil spill?
   - **F** The beach became too slippery to walk on.
   - **G** Many people in Alaska had no oil for their cars.
   - **H** Exxon had to build a new tanker.
   - **I** Many Alaskan fishers lost their businesses.

Passage 2 Whales, dolphins, and porpoises are mammals that belong to the order Cetacea (suh TAY shuh). Cetaceans live throughout the global ocean. They have fishlike bodies and forelimbs called *flippers*. Cetaceans lack hind limbs but have broad, flat tails that help them swim through the water. Cetaceans breathe through blowholes located on the top of the head. They are completely hairless except for a few bristles on their snout. A thick layer of blubber below the skin helps insulate cetaceans against cold temperatures. Cetaceans are divided into two groups: toothed whales and baleen whales. Toothed whales include sperm whales, beluga whales, narwhals, killer whales, dolphins, and porpoises. Baleen whales, such as blue whales, lack teeth. They filter food from the water by using a meshlike net of baleen that hangs from the roof of their mouth.

1. How are organisms that make up the order Cetacea divided?
   - **A** They are divided into cetaceans that have hair and cetaceans that do not have hair.
   - **B** They are divided into cetaceans that have flippers and cetaceans that do not have flippers.
   - **C** They are divided into cetaceans that have blowholes and cetaceans that do not have blowholes.
   - **D** They are divided into cetaceans that have teeth and cetaceans that do not have teeth.

2. Which of the following statements lists characteristics of all cetaceans?
   - **F** Cetaceans have fur and claws and live in rivers.
   - **G** Cetaceans have flippers and bristles on the snout and live in oceans.
   - **H** Cetaceans have blowholes and flippers and live in lakes.
   - **I** Cetaceans have fur, bristles on the snout, and flippers.

Passage 2

1. D
2. G

TEST DOCTOR

Question 1: Answer A is incorrect because according to the passage, cetaceans are completely hairless. Answer B is incorrect because according to the passage, all cetaceans have flippers. Answer C is incorrect because according to the passage, all cetaceans breathe through blowholes located on the top of the head. Answer D is correct because the fact is mentioned in the passage that cetaceans are divided into two groups: toothed whales and baleen whales. The passage states that baleen whales lack teeth.

INTERPRETING GRAPHICS

Use the image of the ocean floor below to answer the questions that follow.

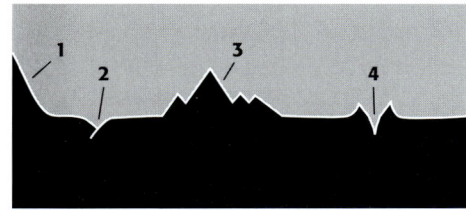

1. At which point are two tectonic plates separating?

A 1

B 2

C 3

D 4

2. Which point shows an ocean trench?

F 1

G 2

H 3

I 4

3. Which feature might eventually become a volcanic island?

A 1

B 2

C 3

D 4

4. Which features are part of the deep-ocean basin?

F 2, 3, and 4

G 1, 2, and 3

H 1, 3, and 4

I 1, 2, and 4

5. Which feature is part of the continental margin?

A 1

B 2

C 3

D 4

MATH

Read each question below, and choose the best answer.

1. Imagine that you are in the kelp-farming business and your kelp grows 33 cm per day. You begin harvesting when your plants are 50 cm tall. During the first 7 days of harvest, you cut 10 cm off the top of your kelp plants each day. How tall will your kelp plants be after the seventh day of harvesting?

A 80 cm

B 130 cm

C 210 cm

D 211 cm

2. A sample of ocean water contains 36 g of dissolved solids per 1,000 g of water. So, how many grams of dissolved solids will be in 4 kg of ocean water?

F 36,000 g

G 360 g

H 250 g

I 144 g

3. If the average depth of the Pacific Ocean is 4,250 m and the average depth of the Atlantic Ocean is 4,000 m, what is the average depth of the two oceans?

A 4,250 m

B 4,150 m

C 4,125 m

D 4,000 m

4. *Alvin,* a minisub, starts at −300 m, then rises 20 m, then drops 150 m, and finally reaches the ocean floor by dropping another 218 m. At what depth is *Alvin* when it reaches the ocean floor?

F −648 m

G −88 m

H 88 m

I 648 m

5. The speed of sound in water is 1,500 m/s. How far will sound travel in water in 1 min?

A 25 m

B 1,500 m

C 9,000 m

D 90,000 m

INTERPRETING GRAPHICS

1. D

2. G

3. C

4. F

5. A

+ TEST DOCTOR

Question 1: Answer D is correct because number 4 is pointing to a mid-ocean ridge. Mid-ocean ridges are mountain chains that form where tectonic plates pull apart and create a series of cracks in the ocean floor. Number 4 is the only part of the image that shows a separation in the ocean floor.

Question 5: Answer A is correct because number 1 is pointing to the continental slope of the ocean floor. The continental slope is one division of the continental margin. Answers B through D are all features of the deep-ocean basin.

MATH

1. D

2. I

3. C

4. F

5. D

+ TEST DOCTOR

Question 2: Students may arrive at the incorrect answer F by multiplying 36 by 1,000 instead of multiplying 36 by 4. Students may arrive at the incorrect answer H by dividing 1,000 by 4 instead of multiplying 36 by 4. Answer I is correct because if there is 36 g of dissolved solids in each kilogram of ocean water, then 4 kg of ocean water should contain 144 g of dissolved solids (36 × 4 = 144).

Scientific Discoveries

Background

According to accounts, the giant squid can put up quite a fight against a whale. Lighthouse keepers in South Africa claim to have seen a giant squid attack and subsequently drown a baby southern whale after an intense battle that lasted for more than an hour.

Researchers do know that squids have excellent eyesight and some of the largest eyes in the animal kingdom. They also have one of the most highly developed brains of any invertebrate.

Science, Technology, and Society

Background

Artificial reefs are usually constructed of materials that are environmentally friendly. The most ideal materials are concrete and steel. It is important to use materials that do not break down and cause pollution. Old ships and planes make ideal artificial reefs because of the materials from which they are constructed. Before the plane or ship is sunk to make a reef, all fuel and oil are removed so that they do not leak out and pollute the area. Other structures that are good for artificial reefs are retired oil rigs and pieces of old concrete foundations.

Science in Action

Scientific Discoveries

In Search of the Giant Squid

You might think that giant squids exist only in science fiction novels. You aren't alone, because many people have never seen a giant squid or do not know that giant squids exist. Scientists have not been able to study giant squids in the ocean. They have been able to study only dead or dying squids that have washed ashore or that have been trapped in fishing nets. As the largest of all invertebrates, giant squids range from 8 to 25 m long and have a mass of as much as 2,000 kg. Giant squids are very similar to smaller squids. But a giant squid's body parts are much larger. For example, a giant squid's eye may be as large as a volleyball! Because of the size of giant squids, you may think that they don't have any enemies in the ocean, but they do. They are usually eaten by sperm whales that can weigh 20 tons!

Math ACTIVITY

A giant squid that washed ashore has a mass of 900 kg. A deep-sea squid that washed ashore has a mass that is 93% smaller than the mass of the giant squid. What is the mass in kilograms of the deep-sea squid?

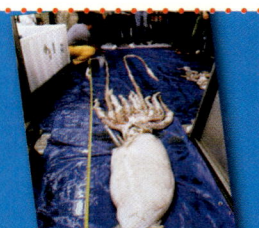

Science, Technology, and Society

Creating Artificial Reefs

If you found a sunken ship, would you look for hidden treasure? Treasure is not the only thing that sunken ships are known for. Hundreds of years ago, people found that the fishing is often good over a sunken ship. The fishing is good because many marine organisms, such as seaweed, corals, and oysters, live only where they can attach to a hard surface in clear water. They attract other organisms to the sunken ship and eventually form a reef community. Thus, in recent years, many human communities have created artificial reefs by sinking objects such as warships, barges, concrete, airplanes, and school buses in the ocean. Like natural reefs, artificial reefs provide a home for organisms and protect organisms from predators.

Social Studies ACTIVITY

WRITING SKILL Research how some artificial reefs are created off the coast of some states in the United States. Write a report that describes some of the objects used to create artificial reefs. In your report also include what countries other than the United States create artificial reefs and what are the benefits and disadvantages of creating artificial reefs.

Answer to Math Activity

(900 kg × 0.93) = 837 kg,
(900 kg − 837 kg) = 63 kg

Answer to Social Studies Activity

Suggest that students research library resources or the Internet to answer the questions. Artificial reefs benefit organisms by giving them a place to live, feed, and breed. Artificial reefs may be beneficial in any aquatic ecosystem because they can replace lost habitat that is needed by the organisms living there. Artificial reefs are not entirely problem free. Artificial reefs must be built safely so that they do not interfere with the movement of ships, they do not entangle divers, and do not pollute the water or fall apart.

Jacques Cousteau

Ocean Explorer Jacques Cousteau was born in France in 1910. Cousteau performed his first underwater diving mission at age 10 and became very fascinated with the possibilities of seeing and breathing underwater. As a result, in 1943, Cousteau and Emile Gagnan developed the first aqualung, a self-contained breathing system for underwater exploration. Using the aqualung and other underwater equipment that he developed, Cousteau began making underwater films. In 1950, Cousteau transformed the *Calypso*, a retired minesweeper boat, into an oceanographic vessel and laboratory. For the next 40 years, Cousteau sailed with the *Calypso* around the world to explore and film the world's oceans. Cousteau produced more than 115 films, many of which have won awards.

Jacques Cousteau opened the eyes of countless people to the sea. During his long life, Cousteau explored Earth's oceans and documented the amazing variety of life that they contain. He was an environmentalist, inventor, and teacher who inspired millions with his joy and wonder of the ocean. Cousteau was an outspoken defender of the environment. He campaigned vigorously to protect the oceans and environment. Cousteau died in 1997 at age 87. Before his death, he dedicated the *Calypso II*, a new research vessel, to the children of the world.

Language Arts ACTIVITY

 Ocean pollution and overfishing are subjects of intense debate. Think about these issues, and discuss them with your classmates. Take notes on what you discuss with your classmates. Then, write an essay in which you try to convince readers of your point of view.

Cousteau sailed his ship, the Calypso, around the world exploring and filming the world's oceans.

To learn more about these Science in Action topics, visit **go.hrw.com** and type in the keyword **HZ5OCEF**.

Current Science
Check out Current Science® articles related to this chapter by visiting **go.hrw.com**. Just type in the keyword **HZ5CS13**.

The Movement of Ocean Water
Chapter Planning Guide

Compression guide:
To shorten instruction because of time limitations, omit the Chapter Lab.

OBJECTIVES	LABS, DEMONSTRATIONS, AND ACTIVITIES	TECHNOLOGY RESOURCES
PACING • 90 min pp. 78–85 **Chapter Opener**	SE **Start-up Activity**, p. 79 ◆ GENERAL	OSP **Parent Letter** ■ GENERAL CD **Student Edition on CD-ROM** CD **Guided Reading Audio CD** ■ TR **Chapter Starter Transparency*** VID **Brain Food Video Quiz**
Section 1 Currents • Describe surface currents. • List the three factors that control surface currents. • Describe deep currents. • Identify the three factors that form deep currents.	SE **School-to-Home Activity** Coriolis Effect in Your Sink?, p. 82 GENERAL TE **Activity** The Coriolis Effect ◆ BASIC TE **Connection Activity** Life Science, p. 81 ADVANCED TE **Activity** Differences in Currents, p. 83 BASIC TE **Connection Activity** Math, p. 85 GENERAL SE **Skills Practice Lab** Up From the Depths, p. 100 ◆ GENERAL CRF **Datasheet for Chapter Lab*** LB **Whiz-Bang Demonstrations** Spin Cycle* ◆ ADVANCED	CRF **Lesson Plans*** TR **Bellringer Transparency*** TR **Earth's Surface Currents*** TR **How Deep Currents Form*** CRF **SciLinks Activity*** GENERAL CD **Interactive Explorations CD-ROM** Latitude Attitude GENERAL VID **Lab Videos for Earth Science** CD **Science Tutor**
PACING • 45 min pp. 86–89 **Section 2 Currents and Climate** • Explain how currents affect climate. • Describe the effects of El Niño. • Explain how scientists study and predict the pattern of El Niño.	TE **Activity** Graphing Temperatures, p. 87 GENERAL SE **Connection to Environmental Science** El Niño and Coral Reefs, p. 88 GENERAL TE **Connection Activity** Real World, p. 88 GENERAL	SE **Internet Activity**, p. 89 GENERAL CRF **Lesson Plans*** TR **Bellringer Transparency*** TR **Upwelling*** CD **Science Tutor**
PACING • 45 min pp. 90–95 **Section 3 Waves** • Identify the parts of a wave. • Explain how the parts of a wave relate to wave movement. • Describe how ocean waves form and move. • Classify types of waves.	TE **Demonstration** Making Waves, p. 90 ◆ GENERAL TE **Group Activity** Modeling Waves, p. 91 BASIC TE **Connection Activity** Math, p. 91 GENERAL SE **Quick Lab** Doing the Wave, p. 92 GENERAL CRF **Datasheet for Quick Lab*** TE **Connection Activity** Real World, p. 92 GENERAL TE **Connection Activity** Math, p. 93 GENERAL	CRF **Lesson Plans*** TR **Bellringer Transparency*** TR **How Deep-Water Waves Become Shallow-Water Waves*** TR **LINK TO PHYSICAL SCIENCE** Measuring Wavelength; Measuring Frequency* CD **Science Tutor**
PACING • 45 min pp. 96–99 **Section 4 Tides** • Explain tides and their relationship with the Earth, sun, and moon. • Describe four different types of tides. • Analyze the relationship between tides and coastal land.	TE **Activity** Intertidal Zones, p. 96 ◆ GENERAL SE **Model-Making Lab** Turning the Tides, p. 114 GENERAL CRF **Datasheet for LabBook*** LB **Long-Term Projects & Research Ideas** An Ocean Commotion* ADVANCED SE **Science in Action** Math, Social Studies, and Language Arts Activities, pp. 106–107 GENERAL	CRF **Lesson Plans*** TR **Bellringer Transparency*** TR **Tidal Variations: Spring Tides; Neap Tides*** CD **Science Tutor**

PACING • 90 min

CHAPTER REVIEW, ASSESSMENT, AND STANDARDIZED TEST PREPARATION

CRF **Vocabulary Activity*** GENERAL
SE **Chapter Review**, pp. 102–103 GENERAL
CRF **Chapter Review*** ■ GENERAL
CRF **Chapter Tests A*** ■ GENERAL, B* ADVANCED, C* SPECIAL NEEDS
SE **Standardized Test Preparation**, pp. 104–105 GENERAL
CRF **Standardized Test Preparation*** GENERAL
CRF **Performance-Based Assessment*** GENERAL
OSP **Test Generator** GENERAL
CRF **Test Item Listing*** GENERAL

Online and Technology Resources

Visit **go.hrw.com** for a variety of free resources related to this textbook. Enter the keyword **HZ5H2O**.

Students can access interactive problem-solving help and active visual concept development with the *Holt Science and Technology* Online Edition available at **www.hrw.com**.

 Guided Reading Audio CD
Also in Spanish

A direct reading of each chapter for auditory learners, reluctant readers, and Spanish-speaking students.

 Science Tutor CD-ROM

Excellent for remediation and test practice.

SKILLS DEVELOPMENT RESOURCES	SECTION REVIEW AND ASSESSMENT	STANDARDS CORRELATIONS
SE Pre-Reading Activity, p. 78 `GENERAL` **OSP** Science Puzzlers, Twisters & Teasers `GENERAL`		National Science Education Standards UCP 1, 2; SAI 1
CRF Directed Reading A* ■ `BASIC`, B* `SPECIAL NEEDS` **CRF** Vocabulary and Section Summary* ■ `GENERAL` **SE** Reading Strategy Reading Organizer, p. 80 `GENERAL` **TE** Reading Strategy Prediction Guide, p. 81 `GENERAL` **TE** Inclusion Strategies, p. 81 ◆ **MS** Math Skills for Science What Is Scientific Notation?* `GENERAL` **SE** Connection to Physics Convection Currents, p. 83 `GENERAL` **TE** Connection to Language Arts Endurance: Shackleton's Incredible Voyage, p. 83 `ADVANCED`	**SE** Reading Checks, pp. 80, 82, 83, 84 `GENERAL` **TE** Homework, p. 83 `GENERAL` **TE** Reteaching, p. 84 `BASIC` **TE** Quiz, p. 84 `GENERAL` **TE** Alternative Assessment, p. 84 `GENERAL` **SE** Section Review,* p. 85 `GENERAL` **CRF** Section Quiz* ■ `GENERAL`	UCP 2; SAI 1; HNS 1, 3; ES 1j; *Chapter Lab:* UCP 2; SAI 1
CRF Directed Reading A* ■ `BASIC`, B* `SPECIAL NEEDS` **CRF** Vocabulary and Section Summary* ■ `GENERAL` **SE** Reading Strategy Paired Summarizing, p. 86 `GENERAL` **TE** Connection to History Benjamin Franklin's Navigation Charts, p. 86 `GENERAL`	**SE** Reading Checks, pp. 87, 89 `GENERAL` **TE** Reteaching, p. 88 `BASIC` **TE** Quiz, p. 88 `GENERAL` **TE** Alternative Assessment, p. 88 `GENERAL` **SE** Section Review,* p. 89 `GENERAL` **CRF** Section Quiz* ■ `GENERAL`	UCP 2; SAI 1; ES 1j
CRF Directed Reading A* ■ `BASIC`, B* `SPECIAL NEEDS` **CRF** Vocabulary and Section Summary* ■ `GENERAL` **SE** Reading Strategy Prediction Guide, p. 90 `GENERAL` **TE** Inclusion Strategies, p. 91 ◆ **TE** Reading Strategy Types of Waves, p. 92 `BASIC` **MS** Math Skills for Science What is a Fraction?* `GENERAL` **TE** Connection to Environmental Science Beach Nourishment, p. 93 `GENERAL` **TE** Connection to Meteorology NOAA, p. 94 `ADVANCED` **CRF** Reinforcement Worksheet Waves to Your Pen Pal* `BASIC`	**SE** Reading Checks, pp. 90, 92, 95 `GENERAL` **TE** Homework, p. 93 `ADVANCED` **TE** Reteaching, p. 94 `BASIC` **TE** Quiz, p. 94 `GENERAL` **TE** Alternative Assessment, p. 94 `GENERAL` **TE** Homework, p. 94 `GENERAL` **SE** Section Review,* p. 95 ■ `GENERAL` **CRF** Section Quiz* ■ `GENERAL`	UCP 1, 2, 3; SAI 1; SPSP 3, 4; ES 1b
CRF Directed Reading A* ■ `BASIC`, B* `SPECIAL NEEDS` **CRF** Vocabulary and Section Summary* ■ `GENERAL` **SE** Reading Strategy Discussion, p. 96 `GENERAL` **SE** Connection to Language Arts Mont-St-Michel Is Sometimes an Island?, p. 97 `GENERAL` **CRF** Reinforcement Worksheet But What About the Tides?* `BASIC` **CRF** Critical Thinking Tides of Trouble* `ADVANCED`	**SE** Reading Checks, pp. 96, 98 `GENERAL` **TE** Reteaching, p. 98 `BASIC` **TE** Quiz, p. 98 `GENERAL` **TE** Alternative Assessment, p. 98 `GENERAL` **TE** Homework, p. 98 `GENERAL` **SE** Section Review,* p. 99 ■ `GENERAL` **CRF** Section Quiz* ■ `GENERAL`	SAI 1; ST 1; HNS 1, 3; ES 3c; *LabBook:* UCP 2; SAI 1; ST 1; ES 3c

One-Stop
Planner® CD-ROM

This convenient CD-ROM includes:
- Lab Materials QuickList Software
- Holt Calendar Planner
- Customizable Lesson Plans
- Printable Worksheets
- ExamView® Test Generator

cnnstudentnews.com

Find the latest news, lesson plans, and activities related to important scientific events.

www.scilinks.org

Maintained by the **National Science Teachers Association.** See Chapter Enrichment pages for a complete list of topics.

Check out *Current Science* articles and activities by visiting the HRW Web site at **go.hrw.com.** Just type in the keyword **HZ5CS14T.**

Classroom Videos

- **Lab Videos** demonstrate the chapter lab.
- **Brain Food Video Quizzes** help students review the chapter material.
- **CNN Videos** bring science into your students' daily life.

Chapter Resources

Visual Resources

CHAPTER STARTER TRANSPARENCY

BELLRINGER TRANSPARENCIES

TEACHING TRANSPARENCIES

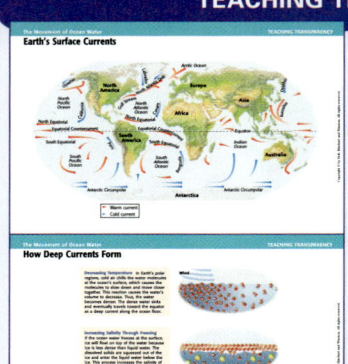

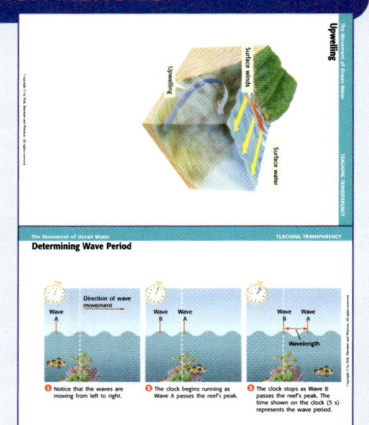

TEACHING TRANSPARENCIES

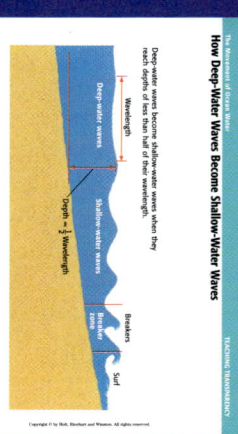

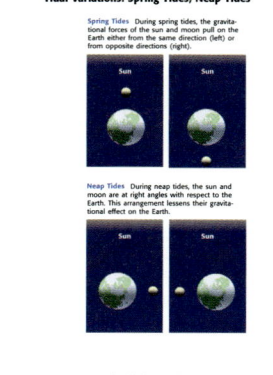

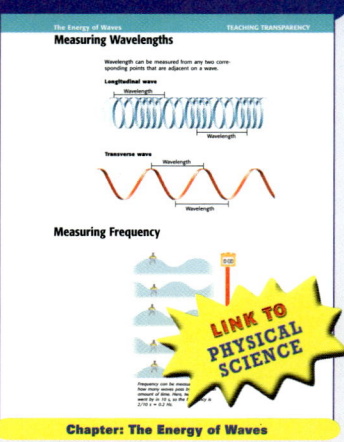

CONCEPT MAPPING TRANSPARENCY

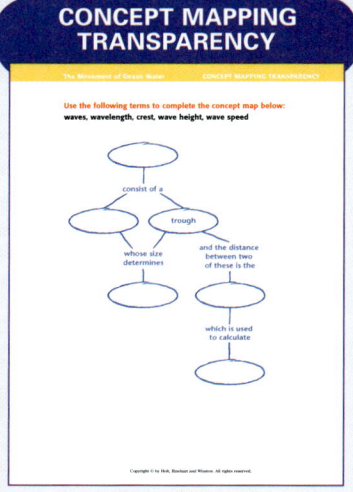

Planning Resources

LESSON PLANS

PARENT LETTER

ALSO IN SPANISH

TEST ITEM LISTING

One-Stop Planner® CD-ROM

This CD-ROM includes all of the resources shown here and the following time-saving tools:

- *Lab Materials QuickList Software*
- *Customizable lesson plans*
- *Holt Calendar Planner*
- *The powerful ExamView® Test Generator*

Meeting Individual Needs

DIRECTED READING A
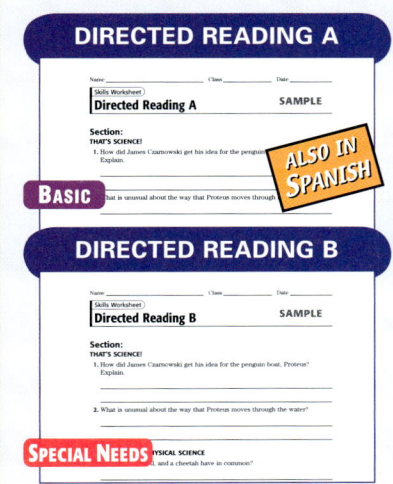
BASIC · ALSO IN SPANISH

VOCABULARY ACTIVITY
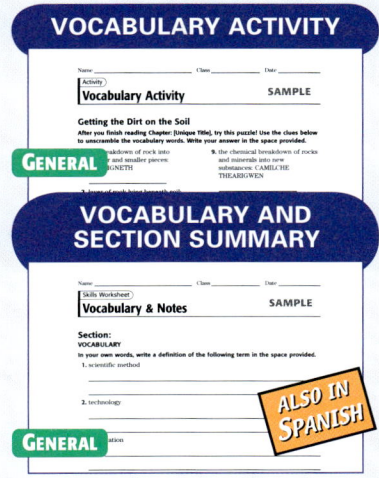
GENERAL

REINFORCEMENT
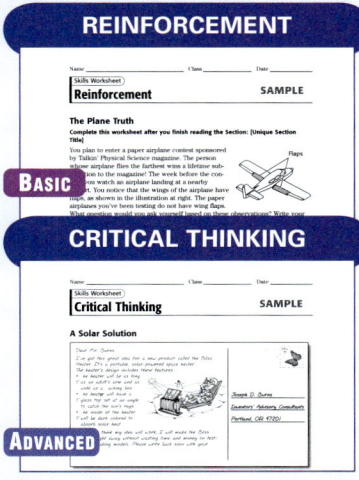
BASIC

SCILINKS ACTIVITY
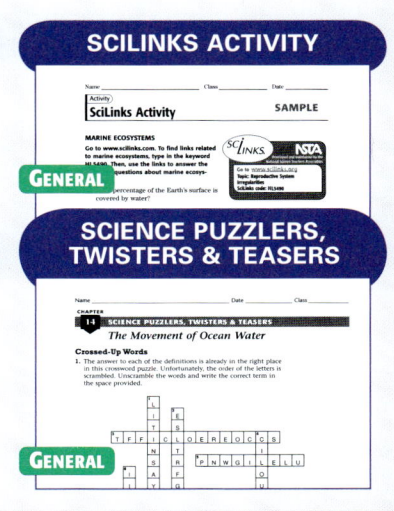
GENERAL

DIRECTED READING B
SPECIAL NEEDS

VOCABULARY AND SECTION SUMMARY
ALSO IN SPANISH
GENERAL

CRITICAL THINKING
ADVANCED

SCIENCE PUZZLERS, TWISTERS & TEASERS
GENERAL

Labs and Activities

LONG-TERM PROJECTS & RESEARCH IDEAS
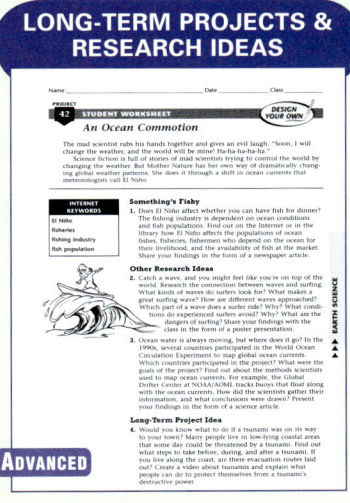
ADVANCED

WHIZ-BANG DEMONSTRATIONS
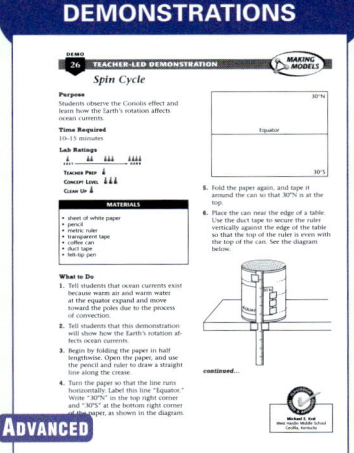
ADVANCED

DATASHEETS FOR QUICKLABS
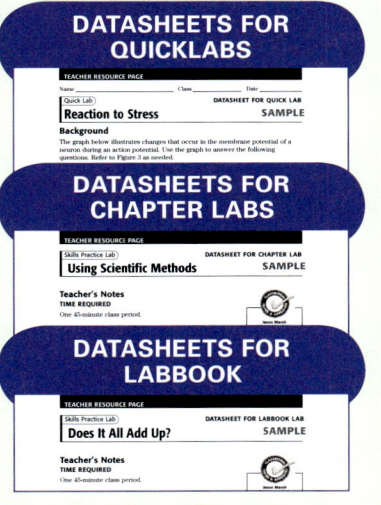

DATASHEETS FOR CHAPTER LABS

DATASHEETS FOR LABBOOK

Review and Assessments

SECTION QUIZ
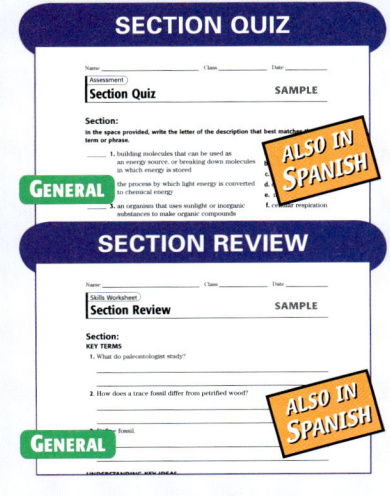
ALSO IN SPANISH
GENERAL

CHAPTER REVIEW
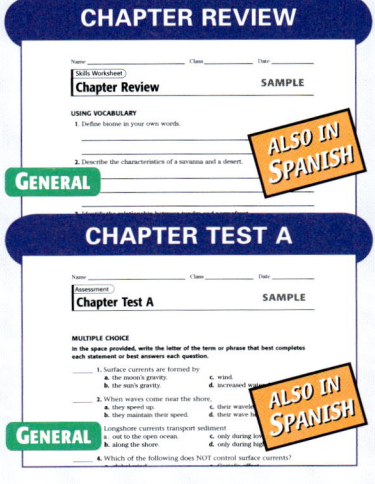
ALSO IN SPANISH
GENERAL

CHAPTER TEST B
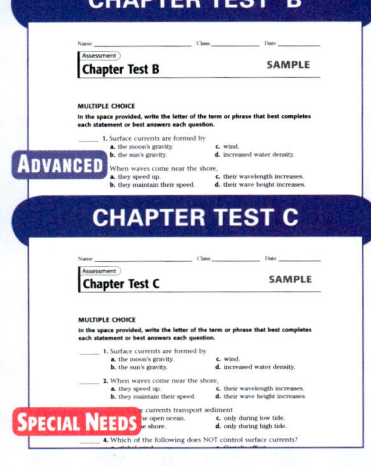
ADVANCED

STANDARDIZED TEST PREPARATION
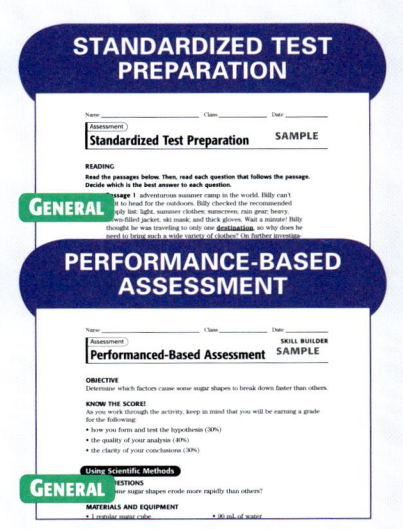
GENERAL

SECTION REVIEW
ALSO IN SPANISH
GENERAL

CHAPTER TEST A
ALSO IN SPANISH
GENERAL

CHAPTER TEST C
SPECIAL NEEDS

PERFORMANCE-BASED ASSESSMENT
GENERAL

This Chapter Enrichment provides relevant and interesting information to expand and enhance your presentation of the chapter material.

Section 1

Currents

Solar Radiation

● One of the fundamental energy sources for all ocean currents is solar radiation. Uneven heating of the Earth by the sun creates differences in air pressure. These differences create wind, which drives surface currents. The sun's energy also creates temperature differences in ocean water, driving deep currents.

The Ra II Expedition

● On May 17, 1970, Thor Heyerdahl's *Ra II* expedition attempted to demonstrate that mariners from ancient Egypt could have reached the New World. The eight-man expedition successfully reached its destination in Barbados on July 12, 1970.

Is That a Fact!

◆ In addition to the westward surface currents that form along the equator, a strong eastward current forms. This current, called the *equatorial counter-current,* flows alongside or just beneath the westward equatorial currents at depths of up to 100 m.

Is That a Fact!

◆ The strongest and largest ocean current is the Antarctic Circumpolar Current, which is estimated to flow at a rate of 125 million cubic meters per second.

◆ One of the fastest ocean currents is the Somali Current, in the western Indian Ocean, which flows at a speed of 14.5 km/h.

◆ The Weddell Sea, where the Antarctic Bottom Water is thought to originate, has the clearest water of any sea. Its clarity has been recorded to a depth of nearly 80 m. In other words, water collected from the upper 80 m of the Weddell Sea is as clear as you would find in a glass of distilled water.

Section 2

Currents and Climate

Global Weather Effects of El Niño

● El Niño affects almost every region of the world. El Niño can cause flooding, landslides, erosion, and drought. Areas that receive excessive moisture, such as the coastal regions of Ecuador and northern Peru, can experience infestations of insects. California usually experiences heavy rainfall throughout the winter. The storms associated with the 1997–1998 El Niño cost northern California more than $150 million in landslide damage. Southeast Asia usually experiences drought and forest fires, which affect the agricultural industry.

Section 3

Waves

Swells

● Swells are generated in the open ocean by wind and can travel thousands of kilometers to shore. These long-wavelength waves have periods of 10 to 30 s. When swells reach shallower water, their height increases, so they fall forward. When this occurs, the waves are called *breakers.*

Tsunamis

- *Tsunami* is a Japanese word that means "harbor wave." Japan has experienced many devastating tsunamis throughout history. The subduction of tectonic plates off the coast of Japan generates the seismic energy necessary to cause tsunamis. Because the Japanese islands are on the edge of deep water and the coastline is rugged with many small harbors, tsunamis have been particularly destructive. The deep water close to the Japanese islands keeps tsunami wave heights short until the waves are very close to shore; when the waves enter shallow water, they suddenly grow taller. The narrow shape of many Japanese harbors causes tsunamis to grow even taller.

Is That a Fact!

- Tsunamis have the potential to be the most destructive ocean waves. Their speed averages 500 km/h, and their period ranges from 5 to 60 min. Because the wave height of a tsunami is usually less than 2 m in the open ocean, they often pass unnoticed beneath ships.

- The earthquakes that produce destructive tsunamis are generally greater than 6.5 on the Richter scale. Most occur in the Pacific Ocean, where there is a high level of seismic activity near plate boundaries.

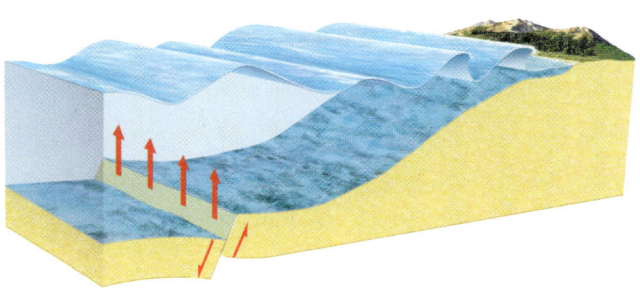

Section 4

Tides

The Bay of Fundy

- The Bay of Fundy experiences the greatest tidal range in the world. The Bay of Fundy is located between New Brunswick and Nova Scotia in Canada and has an average tidal range of 12 m. The primary cause of the extreme tidal range is the Bay of Fundy–Gulf of Maine system. The Atlantic tides push water into the Bay of Fundy–Gulf of Maine basin, which causes the large back-and-forth motion of the tides. A tiny portion of energy from the tides at the Bay of Fundy is being converted into commercial electrical energy.

Is That a Fact!

- The sun exerts only about half the tidal force on Earth that the moon does. The reason is that the distance between Earth and the sun is much greater than the distance between Earth and the moon.

- The Great Lakes also experience tides. Although the tidal range is smaller than that of the oceans.

SciLinks is maintained by the National Science Teachers Association

Developed and maintained by the
National Science Teachers Association

SciLinks is maintained by the National Science Teachers Association to provide you and your students with interesting, up-to-date links that will enrich your classroom presentation of the chapter.

Visit www.scilinks.org and enter the SciLinks code for more information about the topic listed.

Topic: Ocean Currents
SciLinks code: HSM1061

Topic: Ocean Waves
SciLinks code: HSM1066

Topic: El Niño
SciLinks code: HSM0468

Topic: Tides
SciLinks code: HSM1525

Overview

Tell students that this chapter will help them learn about the different factors that affect the movement of ocean water. The chapter describes the different movements of ocean water, including currents, waves, and tides. It also describes how these movements affect land, climate, and organisms.

Assessing Prior Knowledge

Students should be familiar with the following topics:

- ocean-floor topography
- plate tectonics

Identifying Misconceptions

Students tend to confuse the movement of a wave with the movement of the medium (water). Tell students that waves travel through water, the water itself remains basically stationary. You may want to demonstrate this phenomenon using a string tied to a doorknob. Have students snap the string to create a wave that travels the length of the string. Explain that the wave moves from the hand to the doorknob, but the string does not move horizontally.

The Movement of Ocean Water

About the

No, this isn't a traffic jam or the result of careless navigation. Hurricane Hugo is to blame for this major boat pile up. When Hurricane Hugo hit South Carolina's coast in 1989, the hurricane's strong winds created large ocean waves. These ocean waves carried these boats right onto the shore.

PRE-READING ACTIVITY

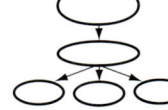 **Concept Map** Before you read the chapter, create the graphic organizer entitled "Concept Map" described in the **Study Skills** section of the Appendix. As you read the chapter, fill in the concept map with details about each type of ocean water movement.

Standards Correlations

National Science Education Standards

The following codes indicate the National Science Education Standards that correlate to this chapter. The full text of the standards is at the front of the book.

Chapter Opener
UCP 1, 2; SAI 1

Section 1 Currents
UCP 2; SAI 1; HNS 1, 3; ES 1j

Section 2 Currents and Climate
UCP 2; SAI 1; ES 1j

Section 3 Waves
UCP 1, 2, 3; SAI 1; SPSP 3, 4; ES 1b

Section 4 Tides
SAI 1; ST 1; HNS 1, 3; ES 3c; *LabBook:* UCP 2; SAI 1; ST 1; ES 3c

Chapter Lab
UCP 2; SAI 1

Chapter Review
SAI 1; ST 2; SPSP 2, 5; HNS 1; ES 1b, 1j, 3c

Science in Action
SAI 1; ST 2; SPSP 5; HNS 1

START-UP ACTIVITY
MATERIALS

FOR EACH GROUP
- food coloring, blue
- food coloring, red
- pencils (2)
- tub, large
- water

Teacher's Notes: To avoid spilling water after the tub is filled, put the tub in position before adding water. Although food coloring is nontoxic, students should be careful not to spill the food coloring on their skin or clothes.

A large cake pan can be used for this activity. Water depth should be about 5 cm. Advise students to observe the experiment closely—the desired result happens quickly and lasts only a few seconds. Explore what happens with different sizes and shapes of tubs.

Answers

1. Sample answer: When the blue water left the current in which it was circulating, it crossed the middle of the tub and began circling in the opposite direction. The red water followed the same pattern.

2. Answers may vary. Currents circulate in the Northern and Southern Hemispheres. When a current crosses the equator, it joins other currents and eventually circulates in the direction that they are moving.

START-UP ACTIVITY

When Whirls Collide

Some ocean currents flow in a clockwise direction, while other ocean currents flow in a counterclockwise direction. Sometimes these currents collide. In this activity, you and your lab partner will demonstrate how two currents flowing in opposite directions affect one another.

Procedure

1. Fill a large **tub** with **water** 5 cm deep.
2. Add **10 drops of red food coloring** to the water at one end of the tub.
3. Add **10 drops of blue food coloring** to the water at the other end of the tub.
4. Using a **pencil**, quickly stir the water at one end of the tub in a clockwise direction while your partner stirs the water at the other end in a counterclockwise direction. Stir both ends for 5 s.
5. Draw what you see happening in the tub immediately after you stop stirring. (Both ends should be swirling.)

Analysis

1. How did the blue water and the red water interact?
2. How does this activity relate to how ocean currents interact?

This Really Happened!

On February 3, 1963, the ocean tanker SS *Marine Sulphur Queen* sailed eastward around the Florida Keys bound for Virginia. Soon after the ship entered the Atlantic, all radio contact with it was lost. Coast Guard rescuers who tried to locate the ship found only life jackets, oil cans, and debris in the area of the tanker's last known position. No survivors were ever found. Several explanations for the Queen's disappearance were offered. The only certainty was that the ship entered the Bermuda Triangle, never to be heard from again.

The Bermuda Triangle is a part of the Atlantic Ocean extending from the Florida coast to Bermuda and Puerto Rico. Many ships have sailed into this area only to vanish. Is there an explanation for this?

Winds blowing southward across the ocean surface in this area sometimes create large, dangerous waves. The Bermuda Triangle also has a reputation for unpredictable weather. Many sailors have been stranded there without wind for their sails. Strong thunderstorms and waterspouts, which are tornadoes on the surface of the ocean, occur with little warning.

The Bermuda Triangle

Chapter Starter Transparency
Use this transparency to help students begin thinking about the factors that make the Bermuda Triangle dangerous.

Focus

Overview

This section discusses the causes and characteristics of surface currents and deep currents.

 Bellringer

Pass out excerpts from Thor Heyerdahl's *Kon-Tiki* (1950) or *The Ra Expeditions* (1971) to students. Display a large map of the world that shows the different ocean currents. After showing students the departure and destination points for Heyerdahl's voyages, have them hypothesize which currents Heyerdahl would have used to reach his destinations. Point out that DNA testing later showed that Polynesians did not originate in Peru. Have students write about how a scientific model can be plausible but incorrect.

Motivate

Discussion ——— GENERAL

Rivers and Surface Currents

Ask students to compare rivers and surface currents. Explain that rivers and surface currents are similar because both are long, moving bodies of water. However, rivers flow because of the pull of gravity, while surface currents are driven by the wind and the rotation of the Earth.
LS Verbal

SECTION 1

READING WARM-UP

Objectives

- Describe surface currents.
- List the three factors that control surface currents.
- Describe deep currents.
- Identify the three factors that form deep currents.

Terms to Learn

ocean current
surface current
Coriolis effect
deep current

READING STRATEGY

Reading Organizer As you read this section, create an outline of the section. Use the headings from the section in your outline.

ocean current a movement of ocean water that follows a regular pattern

Figure 1 The handcrafted Kon-Tiki was made mainly from materials that would have been available to ancient Peruvians.

Currents

Imagine that you are stranded on a desert island. You stuff a distress message into a bottle and throw it into the ocean. Is there any way to predict where your bottle may land?

Actually, there is a way to predict where the bottle will end up. Ocean water contains streamlike movements of water called **ocean currents.** Currents are influenced by a number of factors, including weather, the Earth's rotation, and the position of the continents. With knowledge of ocean currents, people are able to predict where objects in the open ocean will be carried.

One Way to Explore Currents

In the 1940s, a Norwegian explorer named Thor Heyerdahl tried to answer questions about human migration across the ocean. Heyerdahl theorized that the inhabitants of Polynesia originally sailed from Peru on rafts powered only by the wind and ocean currents. In 1947, Heyerdahl and a crew of five people set sail from Peru on a raft, which is shown in **Figure 1.**

On the 97th day of their expedition, Heyerdahl and his crew landed on an island in Polynesia. Currents had carried the raft westward more than 6,000 km across the South Pacific. This landing supported Heyerdahl's theory that ocean currents carried the ancient Peruvians across the Pacific to Polynesia.

✓ **Reading Check** What was Heyerdahl's theory, and how did he prove it? (*See the Appendix for answers to Reading Checks.*)

Answer to Reading Check

Heyerdahl theorized that the inhabitants of Polynesia originally sailed from Peru on rafts powered only by the wind and ocean currents. Heyerdahl proved his theory by sailing from Peru to Polynesia on a raft powered only by wind and ocean currents.

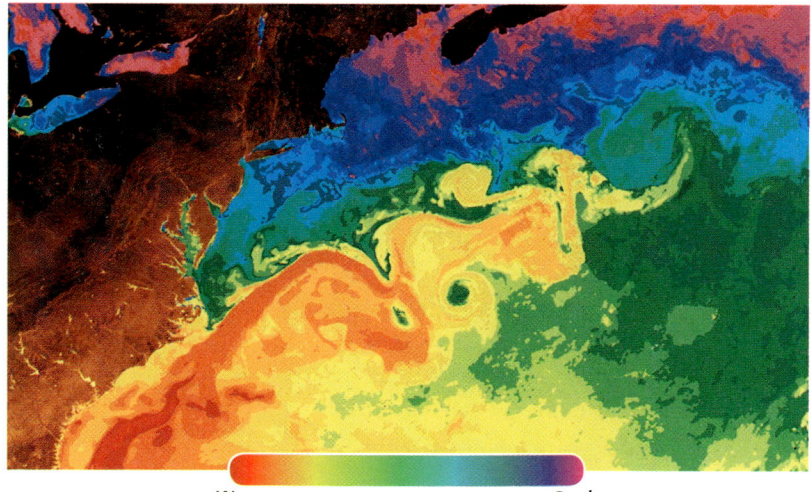

Figure 2 *This infrared satellite image shows the Gulf Stream current moving warm water from lower latitudes to higher latitudes.*

Warm   Cool

Surface Currents

Horizontal, streamlike movements of water that occur at or near the surface of the ocean are called **surface currents.** Surface currents can reach depths of several hundred meters and lengths of several thousand kilometers and can travel across oceans. The Gulf Stream, shown in **Figure 2,** is one of the longest surface currents—it transports 25 times more water than all the rivers in the world.

Surface currents are controlled by three factors: global winds, the Coriolis effect, and continental deflections. These three factors keep surface currents flowing in distinct patterns around the Earth.

Global Winds

Have you ever blown gently on a cup of hot chocolate? You may have noticed ripples moving across the surface, as in **Figure 3.** These ripples are caused by a tiny surface current created by your breath. In much the same way that you create ripples, winds that blow across the Earth's surface create surface currents in the ocean.

Different winds cause currents to flow in different directions. Near the equator, the winds blow ocean water east to west, but closer to the poles, ocean water is blown west to east. Merchant ships often use these currents to travel more quickly back and forth across the oceans.

surface current a horizontal movement of ocean water that is caused by wind and that occurs at or near the ocean's surface

Figure 3 *Winds form surface currents in the ocean, much like blowing on a cup of hot chocolate forms ripples.*

Teach

📖 READING STRATEGY — GENERAL

Prediction Guide Before students read the passage describing the three causes of surface currents, ask them what they think might cause currents on the surface of oceans. Students will discover the answers as they continue reading this section.
LS Logical

CONNECTION ACTiViTY
Life Science ——— ADVANCED

Writing **Current Colonies** Ocean currents help animals and plants colonize new islands. The Galápagos Islands are 1,100 km west of South America. Of the 22 species of reptiles on the islands, 20 are found nowhere else in the world. Scientists speculate that the ancestors of these reptiles found their way from South America to the islands on natural rafts carried by the Humbolt and South Equatorial Currents (the same currents that carried Thor Heyerdahl to Polynesia). Have interested students find out more about the unique plants and animals of the Galápagos Islands. Then, have each student create a poster to showcase his or her research.
LS Visual/Intrapersonal

Is That a Fact!

No metal was used in the construction of the *Kon-Tiki.* The wood raft was made of thick Peruvian balsa logs and featured a bamboo cabin set in the center, a large steering oar at the stern, and five centerboards. Two masts were used to support the rectangular sail. Although it crashed into a reef when it finally reached Polynesia, the *Kon-Tiki* was restored and is currently on display at a museum in Oslo, Norway.

🔴 INCLUSION Strategies

- **Learning Disabled**
- **Attention Deficit Disorder**
- **Hearing Impaired**

Sprinkle pepper on top of the surface of water in a shallow pan. Then, turn on a blow dryer, and blow the pepper across the surface of the water. Organize students into small groups, and ask them to brainstorm how the demonstration is similar to what happens in oceans. Have students share their ideas with the rest of the class. **LS** Visual/Interpersonal

The Coriolis Effect For students having difficulty understanding the Coriolis effect, use a turntable to demonstrate the concept.

1. Cover the entire surface of the turntable with a circle of paper.

2. Spin the turntable platter, and explain that it represents the rotating Earth.

3. Instruct a student to attempt to draw a straight line from the center of the turntable to the edge. A curved line will be formed. The curved line represents the curved path of surface currents due to Earth's rotation.

LS Visual/Kinesthetic   English Language Learners

Answer to Reading Check

The Earth's rotation causes surface currents to move in curved paths rather than in straight lines.

Answer to School-to-Home Activity

Students should find that the Coriolis effect does not influence the direction in which water drains from a sink.

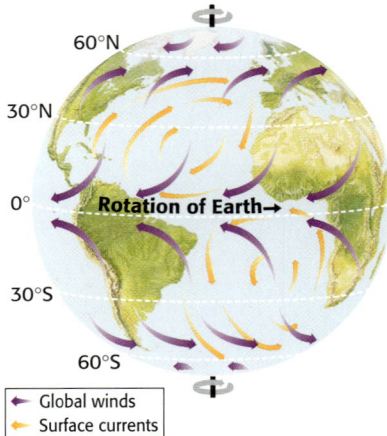

← Global winds
← Surface currents

Figure 4 *The rotation of the Earth causes surface currents (yellow arrows) and global winds (purple arrows) to curve as they move across the Earth's surface.*

Coriolis effect the apparent curving of the path of a moving object from an otherwise straight path due to the Earth's rotation

SCHOOL to HOME

Coriolis Effect in Your Sink?

WRITING SKILL Some people think the Coriolis effect can be seen in sinks. Does water draining from sinks turns clockwise in the Northern Hemisphere and counterclockwise in the Southern Hemisphere? Research this question at the library, on the Internet, and in your sink at home with a parent. Write what you learn in your **science journal.**

ACTIVITY

The Coriolis Effect

The Earth's rotation causes wind and surface currents to move in curved paths rather than in straight lines. The apparent curving of moving objects from a straight path due to the Earth's rotation is called the Coriolis effect. To understand the Coriolis effect, imagine trying to roll a ball straight across a turning merry-go-round. Because the merry-go-round is spinning, the path of the ball will curve before it reaches the other side. **Figure 4** shows how the Coriolis effect causes surface currents in the Northern Hemisphere to turn clockwise, and surface currents in the Southern Hemisphere to turn counterclockwise.

✓ **Reading Check** What causes currents to move in curved paths instead of straight lines?

Continental Deflections

If the Earth's surface were covered only with water, surface currents would travel freely across the globe in a very uniform pattern. However, you know that water does not cover the entire surface of the Earth. Continents rise above sea level over roughly one-third of the Earth's surface. When surface currents meet continents, the currents *deflect,* or change direction. Notice in **Figure 5** how the Brazil Current deflects southward as it meets the east coast of South America.

Atlantic Ocean

South America

Brazil Current

Figure 5 *If South America were not in the way, the Brazil Current would probably flow farther west.*

Is That a Fact!

The Earth's rotation affects ocean currents directly by causing currents to circle in opposite directions on either side of the equator. Earth's rotation also affects the wind patterns, which in turn drive surface currents.

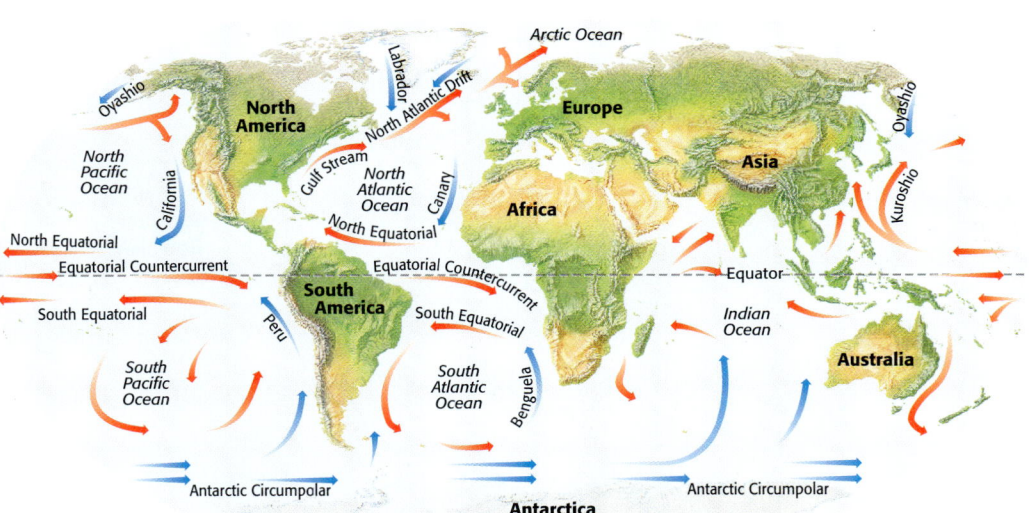

Figure 6 *This map shows Earth's surface currents. Warm-water currents are shown as red arrows, and cold-water currents are shown as blue arrows.*

Warm current
Cold current

ACTIVITY ── BASIC

Differences in Currents Refer to **Figure 6** to point out that the currents in the Southern Hemisphere turn counter-clockwise and the currents in the Northern Hemisphere turn clockwise. Using a globe, show the direction of ocean currents in various areas around the world. Note that a current in the South Atlantic Ocean crosses the equator and joins the clockwise-turning currents. Discuss how this process was modeled in the Start-up Activity at the beginning of this chapter. **English Language Learners**
LS Visual

Homework ── GENERAL

Research Have students choose four coastal cities that are separated by oceans. Ask students to describe in their **science journal** the currents they would use to sail from one city to the next. Students should use **Figure 6** as a reference for determining which currents they would use. For further details on the names of ocean currents, students can use an atlas or another reference tool. They can write their entries in the form of a ship's log.
LS Visual/Intrapersonal

Answer to Reading Check

The three factors that form a pattern of surface currents on Earth are global winds, the Coriolis effect, and continental deflections.

Answer to Connection to Physics

Currents located near the equator carry more thermal energy.

Taking Temperatures

All three factors—global winds, the Coriolis effect, and continental deflections—work together to form a pattern of surface currents on Earth. But currents are also affected by the temperature of the water in which they form. Warm-water currents begin near the equator and carry warm water to other parts of the ocean. Cold-water currents begin closer to the poles and carry cool water to other parts of the ocean. As you can see on the map in **Figure 6,** all the oceans are connected and both warm-water and cold-water currents travel from one ocean to another.

✔ Reading Check What three factors form a pattern of surface currents on Earth?

Deep Currents

Streamlike movements of ocean water located far below the surface are called **deep currents.** Unlike surface currents, deep currents are not directly controlled by wind. Instead, deep currents form in parts of the ocean where water density increases. *Density* is the amount of matter in a given space, or volume. The density of ocean water is affected by temperature and *salinity*—a measure of the amount of dissolved salts or solids in a liquid. Both decreasing the temperature of ocean water and increasing the water's salinity increase the water's density.

deep current a streamlike movement of ocean water far below the surface

CONNECTION TO Physics

Convection Currents While winds are often responsible for ocean currents, the sun is the initial energy source of the winds and currents. Because the sun heats the Earth more in some places than in others, convection currents are formed. These currents transfer thermal energy. Which ocean currents do you think carry more thermal energy, currents located near the equator or currents located near the poles?

CONNECTION to Language Arts ── ADVANCED

Endurance: Shackleton's Incredible Voyage In 1914, British explorer Ernest Shackleton attempted to reach the South Pole in his ship *Endurance*. Just one day's sail from Antarctica, *Endurance* became trapped in sea ice and was frozen for 10 months. Pressure from the ice eventually crushed the ship, and Shackleton and his crew were forced to spend 5 months camping on ice floes. At the end of this 5-month period, Shackleton made a daring attempt to seek help and organize a team to rescue the crew that he was forced to leave behind by crossing 800 miles of open ocean to South Georgia Island. After crossing the mountains of South Georgia, Shackleton was able to reach the island's whaling station. There, he assembled the rescue team that would return to Antarctica to save all of the men he had left behind. This heroic epic is documented in the book *Endurance: Shackleton's Incredible Voyage* by Alfred Lansing, which is available in a print version, an audio book, and a documentary.

Reteaching — BASIC

Movement of Ocean Currents

Ask students to list the three factors that control surface currents. Then, ask students to list the factors that cause deep currents to form. **LS Verbal**

Quiz — GENERAL

1. Give two characteristics and one example of each type of current. (Sample answer: Surface currents occur at the surface of the ocean and are directly influenced by the wind, continental deflections, and the Coriolis effect. The Gulf Stream is an example of a surface current. Deep currents occur deep in the ocean and are influenced by differences in water density. The Antarctic Bottom Water is an example of a deep current.)

Alternative Assessment — GENERAL

Sailing with the Currents Ask students to imagine that they are planning a voyage around the world. They can choose any route they wish, but they must sail with the currents. Have them map out their selected route and show the names of the currents, their point of origin, and their point of destination. **LS Logical**

Formation and Movement of Deep Currents

The relationship between the density of ocean water and the formation of deep currents is shown in **Figure 7.** Differences in temperature and salinity—and the resulting differences in density—cause variations in the movement of deep currents. For example, the deepest current, the Antarctic Bottom Water, is denser than the North Atlantic Deep Water. Both currents spread out across the ocean floor as they flow toward each other. Because less-dense water always flows on top of denser water, the North Atlantic Deep Water flows on top of the Antarctic Bottom Water when the currents meet, as shown in **Figure 8.**

✓ **Reading Check** How does the density of ocean water affect deep currents?

Figure 7 How Deep Currents Form

Decreasing Temperature In Earth's polar regions, cold air chills the water molecules at the ocean's surface, which causes the molecules to slow down and move closer together. This reaction causes the water's volume to decrease. Thus, the water becomes denser. The dense water sinks and eventually travels toward the equator as a deep current along the ocean floor.

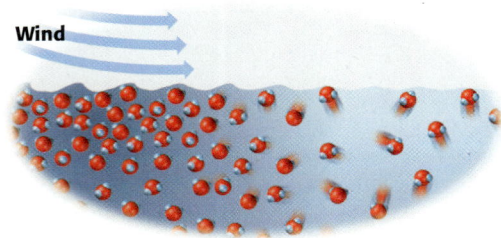
Wind

Increasing Salinity Through Freezing If the ocean water freezes at the surface, ice will float on top of the water because ice is less dense than liquid water. The dissolved solids are squeezed out of the ice and enter the liquid water below the ice. This process increases the salinity of the water. As a result of the increased salinity, the water's density increases.

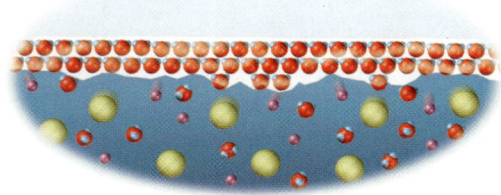

Increasing Salinity Through Evaporation Another way salinity increases is through evaporation of surface water, which removes water but leaves solids behind. This process is especially common in warm climates. Increasing salinity through freezing or evaporation causes water to become denser, to sink to the ocean floor, and to form a deep current.

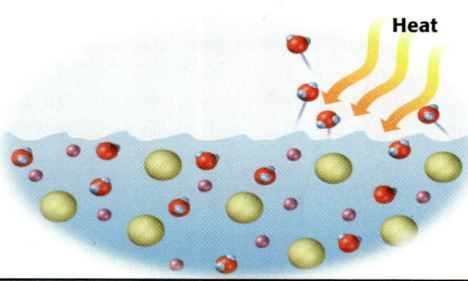
Heat

Answer to Reading Check

Density causes variations in the movement of deep currents.

MISCONCEPTION ALERT

Models of Molecules Remind students that molecules aren't made of little colored balls—the balls in the illustration on this page are models that represent molecules. The red and blue balls that are attached to one another represent water molecules, and the yellow balls represent dissolved solids.

Polar regions

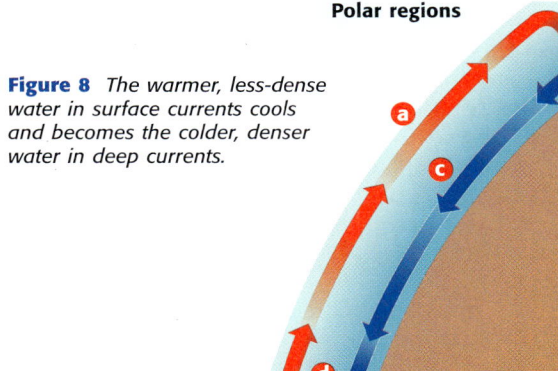

Figure 8 *The warmer, less-dense water in surface currents cools and becomes the colder, denser water in deep currents.*

ⓐ Surface currents carry the warmer, less-dense water from other ocean regions to polar regions.

ⓑ Warm water from surface currents replaces colder, denser water that sinks to the ocean floor.

ⓒ Deep currents carry colder, denser water along the ocean floor from polar regions to other ocean regions.

ⓓ Water from deep currents rises to replace water leaving surface currents.

SECTION Review

Summary

- Surface currents are streamlike movements of water at or near the surface of the ocean.

- Surface currents are controlled by three factors: global winds, the Coriolis effect, and continental deflections.

- Deep currents are streamlike movements of ocean water located far below the surface.

- Deep currents form where the density of ocean water increases. Water density depends on temperature and salinity.

Using Key Terms

The statements below are false. For each statement, replace the underlined word to make a true statement.

1. <u>Deep currents</u> are directly controlled by wind.

2. An increase in density in parts of the ocean can cause <u>surface currents</u> to form.

Understanding Key Ideas

3. Surface currents
  a. are formed by wind.
  b. are streamlike movements of water.
  c. can travel across entire oceans.
  d. All of the above

4. List three factors that control surface currents.

5. How does a continent affect the movement of a surface current?

6. Explain how temperature and salinity affect the formation of deep currents.

Math Skills

7. The Gulf Stream flows along the North Carolina coast at 90 million cubic meters per second and at 40 million cubic meters per second when it turns eastward. How much faster is the Gulf Stream flowing along the coast than when it turns eastward?

Critical Thinking

8. Evaluating Conclusions If there were no land on Earth's surface, what would the pattern of surface currents look like? Explain your answer.

9. Making Comparisons Compare the factors that contribute to the formation of surface currents and deep currents.

Developed and maintained by the National Science Teachers Association

For a variety of links related to this chapter, go to www.scilinks.org

Topic: Ocean Currents
SciLinks code: HSM1061

CONNECTION ACTIVITY
Math — GENERAL

Scientific Notation Review scientific notation with students. Tell them that there are approximately 1.7×10^{20} water molecules in one drop of sea water and 4×10^{43} water molecules in the Mediterranean Sea. Have students help you write these two numbers to show how large they are. For reference, tell students that there are approximately 17 times as many molecules in a drop of water as there are insects on Earth. **LS Visual/Logical**

CHAPTER RESOURCES

Chapter Resource File

- Section Quiz GENERAL
- Section Review GENERAL
- Vocabulary and Section Summary GENERAL
- SciLinks Activity GENERAL

Technology

Transparencies
- How Deep Currents Form

Interactive Explorations CD-ROM
- Latitude Attitude GENERAL

Focus

Overview

This section discusses how currents affect climate. Students will learn about the effects of El Niño and ways in which scientists study El Niño.

 **Bellringer**

Find the average yearly temperatures for the Scilly Isles in England and the average yearly temperatures for Newfoundland, Canada, and display them on the board or an overhead projector. Then, show students where those places are located on a map or globe. Ask students to compare the temperatures of the two locations. Then, explain that surface currents contribute to the different temperatures of the two locations. **LS Visual/Verbal**

Motivate

Discussion ——— GENERAL

The Effects of Currents Ask students to think of ways in which currents affect climate. Ask students to think about how climate changes in areas that are close to bodies of water. **LS Verbal**

READING WARM-UP

Objectives
- Explain how currents affect climate.
- Describe the effects of El Niño.
- Explain how scientists study and predict the pattern of El Niño.

Terms to Learn
upwelling
El Niño
La Niña

READING STRATEGY

Paired Summarizing Read this section silently. In pairs, take turns summarizing the material. Stop to discuss ideas that seem confusing.

Currents and Climate

The Scilly Isles in England are located as far north as Newfoundland in northeast Canada. But the Scilly Isles experience warm temperatures almost all year long, while Newfoundland has long winters of frost and snow. How can two places at similar latitudes have completely different climates? This difference in climate is caused by surface currents.

Surface Currents and Climate

Surface currents greatly affect the climate in many parts of the world. Some surface currents warm or cool coastal areas year-round. Other surface currents sometimes change their circulation pattern. Changes in circulation patterns cause changes in atmosphere that affect the climate in many parts of the world.

Warm-Water Currents and Climate

Although surface currents are generally much warmer than deep currents, the temperatures of surface currents do vary. Surface currents are classified as warm-water currents or cold-water currents. Warm-water currents create warmer climates in coastal areas that would otherwise be much cooler. **Figure 1** shows how the Gulf Stream carries warm water from the Tropics to the North Atlantic Ocean. The Gulf Stream flows to the British Isles and creates a relatively mild climate for land at such high latitude. The Gulf Stream is the same current that makes the climate of the Scilly Isles very different from the climate of Newfoundland.

Figure 1 **How Warm-Water Currents Affect Climate**

Warm-water currents, such as the Gulf Stream, can affect the climate of coastal regions.

❷ The Gulf Stream flows to the British Isles and creates a relatively mild climate for land at such a high latitude.

❶ The Gulf Stream carries warm water from the Tropics to the North Atlantic Ocean.

Gulf Stream

CHAPTER RESOURCES

Chapter Resource File

- Lesson Plan
- Directed Reading A **BASIC**
- Directed Reading B **SPECIAL NEEDS**

Technology

Transparencies
- Bellringer
- Upwelling

CONNECTION to
History ——— GENERAL

Benjamin Franklin's Navigation Charts
Early in U.S. history, Benjamin Franklin noticed that mail ships took much longer to travel from England to the United States than from United States to England. He then discovered that ships from England were sailing against the Gulf Stream. Franklin revolutionized ocean navigation by designing charts that helped sailors avoid sailing against major surface currents.

Figure 2   **How Cold-Water Currents Affect Climate**

Cold-water currents, such as the California Current, can affect the climate of coastal regions.

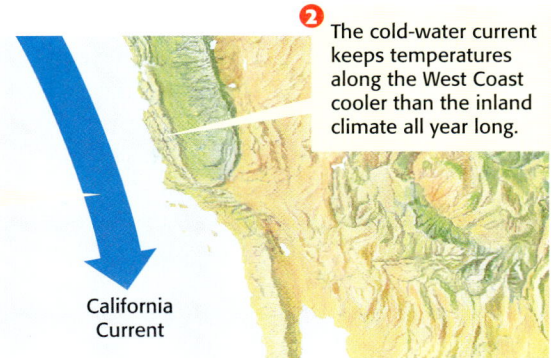

2 The cold-water current keeps temperatures along the West Coast cooler than the inland climate all year long.

1 Cold water from the northern Pacific Ocean is carried south to Mexico by the California Current.

California Current

Cold-Water Currents and Climate

Cold-water currents also affect the climate of the land near where they flow. **Figure 2** shows how the California Current carries cold water from the North Pacific Ocean southward to Mexico. The cold-water California Current keeps the climate along the West Coast cooler than the inland climate year-round.

upwelling the movement of deep, cold, and nutrient-rich water to the surface

✓ **Reading Check** How do cold-water currents affect coastal regions?

Upwelling

When local wind patterns blow along the north-west coast of South America, they cause local surface currents to move away from the shore. This warm water is then replaced by deep, cold water. This movement causes upwelling to occur in the eastern Pacific. **Upwelling** is a process in which cold, nutrient-rich water from the deep ocean rises to the surface and replaces warm surface water, as shown in **Figure 3.** The nutrients from the deep ocean are made up of elements and chemicals, such as iron and nitrate. When these chemicals are brought to the sunny surface, they help tiny plants grow through the process of photosynthesis.

The process of upwelling is extremely important to organisms. The nutrients that are brought to the surface of the ocean support the growth of phytoplankton and zooplankton. These tiny plants and animals support other organisms such as fish and seabirds.

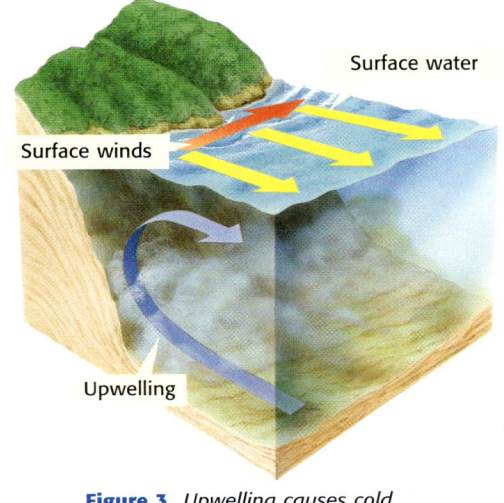

Surface water

Surface winds

Upwelling

Figure 3 Upwelling causes cold, nutrient-rich water from the deep ocean to rise to the surface.

BRAIN FOOD

Surface Currents and Hurricanes

Hurricanes are large tropical storms that originate over large bodies of warm water, such as the Caribbean Sea. Hurricanes are created when warm water builds storm clouds, resulting in a massive low-pressure cell. In North America, hurricanes generally make landfall on the East Coast of the United States or on the coast of the Gulf of Mexico, and they have reached as far north as Maine. On the West Coast, a hurricane has never made landfall in California, because the water off the coast of California is not warm enough to generate storm clouds or a low-pressure cell.

Reteaching — BASIC

Effects of Currents Ask students to describe the effects of currents on climate, land, and organisms. Students can use these notes as a study aid.
LS Logical

Quiz — GENERAL

Ask students whether each of the statements below is true or false.

1. Upwelling is a process in which warm, nutrient-rich water from the deep ocean rises to the surface. (false)

2. El Niño occurs every 2 to 12 years. (true)

3. During an El Niño, California usually experiences a drought. (false)

4. Surface currents greatly affect the climate in many parts of the world. (true)

Alternative Assessment — GENERAL

Studying El Niño Ask students to research how scientists study El Niño. Then, ask students to write a report explaining how the data that scientists collect about El Niño can help prevent future disasters caused by El Niño.
LS Intrapersonal

CONNECTION TO Environmental Science

El Niño and Coral Reefs
The increase of surface water temperatures during El Niño can destroy coral reefs. *Coral reefs* are fragile limestone ridges built by tiny coral animals. An increase in surface water temperature and exposure to the sun (due to a decrease in sea level) can cause corals to become bleached and die. Coral reefs support a diverse community of marine life. Create a world map that shows the locations of the coral reefs.

ACTIVITY

El Niño a change in the water temperature in the Pacific Ocean that produces a warm current

La Niña a change in the eastern Pacific Ocean in which the surface water temperature becomes unusually cool

---

El Niño

Every 2 to 12 years, the South Pacific trade winds move less warm water to the western Pacific than they usually do. Thus, surface-water temperatures along the coast of South America rise. Gradually, this warming spreads westward. This periodic change in the location of warm and cool surface waters in the Pacific Ocean is called **El Niño.** El Niño can last for a year or longer and not only affects the surface waters but also changes the interaction of the ocean and the atmosphere, which in turn changes global weather patterns.

Sometimes, El Niño is followed by La Niña. **La Niña** is a periodic change in the eastern Pacific Ocean in which the surface-water temperature becomes unusually cool. Like El Niño, La Niña also affects weather patterns.

Effects of El Niño

El Niño alters weather patterns enough to cause disasters. These disasters include flash floods and mudslides in areas of the world that usually receive little rain, such as the southern half of the United States and Peru. **Figure 4** shows homes in Southern California destroyed by a mudslide caused by El Niño. While some regions flood, regions that usually get a lot of rain may experience *droughts,* an unusually long period during which rainfall is below average. During El Niño, severe droughts can occur in Indonesia and Australia. Periods of severe drought can lead to crop failure.

During El Niño, the upwelling of nutrient-rich water does not occur off the coast of South America, which affects the organisms that depend on the nutrients for food.

Figure 4 *This damage in Southern California was the result of excessive rain caused by El Niño in 1997.*

CONNECTION ACTIVITY Real World — GENERAL

El Niño El Niño has far-reaching effects on many countries. Many areas have been devastated as a result of El Niño–related floods, storms, and droughts. Have students search for news stories describing some of the effects of El Niño in recent years. Encourage students to contrast these negative effects with the positive effects of El Niño, such as extended growing seasons. **LS Intrapersonal**

Studying and Predicting El Niño

Because El Niño occurs every 2 to 12 years, studying and predicting it can be difficult. However, it is important for scientists to learn as much as possible about El Niño because of its effects on organisms and land.

One way scientists collect data to predict an El Niño is through a network of buoys operated by the National Oceanic and Atmospheric Administration (NOAA). The buoys, some of which are anchored to the ocean floor, are located along the Earth's equator. The buoys record data about surface temperature, air temperature, currents, and winds. The buoys transmit some of the data on a daily basis to NOAA through a satellite in space.

When the buoys report that the South Pacific trade winds are not as strong as they usually are or that the surface temperatures of the tropical oceans have risen, scientists can predict that an El Niño is likely to occur.

Reading Check Why is it important to study El Niño? Describe one way scientists study El Niño.

For another activity related to this chapter, go to go.hrw.com and type in the keyword HZ5H20W.

SECTION Review

Summary

- Surface currents affect the climate of the land near which they flow.
- Warm-water currents bring warmer climates to coastal regions.
- Cold-water currents bring cooler climates to coastal regions.
- During El Niño, warm and cool surface waters change locations.
- El Niño can cause floods, mudslides, and drought.

Using Key Terms

1. Use each of the following terms in a separate sentence: *upwelling*, *El Niño*, and *La Niña*.

Understanding Key Ideas

2. The Gulf Stream carries warm water to the North Atlantic Ocean, which contributes to
   a. a harsh winter in the British Isles.
   b. a cold-water surface current that flows to the British Isles.
   c. a mild climate for the British Isles.
   d. a warm-water surface current that flows along the coast of California.

3. Why might the climate in Scotland be relatively mild even though the country is located at a high latitude?

4. Name two disasters caused by El Niño.

Math Skills

5. A fisher usually catches 540 kg of anchovies off the coast of Peru. During El Niño, the fisher caught 85% less fish. How many kilograms of fish did the fisher catch during El Niño?

Critical Thinking

6. **Applying Concepts** Many marine organisms depend on upwelling to bring nutrients to the surface. How might El Niño affect a fisher's way of life?

For a variety of links related to this chapter, go to www.scilinks.org

Topic: El Niño
SciLinks code: HSM0468

Answers to Section Review

1. Sample answer: Upwelling is a process in which cold, nutrient-rich water from the deep ocean rises to the surface. El Niño can alter weather patterns enough to cause disasters. La Niña can cause surface-water temperatures to become unusually cool.

2. c

3. Even though Scotland is located at a high latitude, its climate is relatively mild because the Gulf Stream carries warm water from the Tropics to the North Atlantic Ocean.

4. El Niño can cause flash floods and landslides.

5. 540 kg × 0.85 = 459 kg;
   540 kg − 459 kg = 81 kg

6. Sample answer: During El Niño, upwelling does not occur along the coast of Peru. When upwelling does not occur, nutrients from deep water do not rise to the surface. Marine organisms that depend on the nutrients for food may die or move to other areas in search of food. Fishers who depend on these organisms may have reduced catch, which will negatively affect fishers' way of life.

Answer to Reading Check

Sample answer: Answers may vary. It is important to study El Niño because El Niño can greatly affect organisms and land. One way that scientists study El Niño is through a network of buoys located along the equator. These buoys record information that helps scientists predict when an El Niño is likely to occur.

CHAPTER RESOURCES

Chapter Resource File

- Section Quiz **GENERAL**
- Section Review **GENERAL**
- Vocabulary and Section Summary **GENERAL**

Focus

Overview

This section describes the characteristics of waves. Students will learn about wave formation and movement, ways to identify different types of waves, and ways to measure different wave features. The section also discusses dangerous movements of ocean water.

🔔 Bellringer

Describe the following scenario to students: "You are floating in the ocean 1 km from shore, which is north of you. A surface current is flowing east. Are you more likely to travel north with the waves toward the shore or east with the surface current?"

(east, because wave energy travels through the water, but the water doesn't travel with the waves)

Motivate

Demonstration —— GENERAL

Making Waves Fill a large tub with water, and place it at the front of the classroom. Ask students to suggest ways to produce waves in the tub without touching the water. Demonstrate the suggestions. Discuss whether each method of producing waves is similar to any of the ways that ocean waves are naturally produced. **LS Visual**

READING WARM-UP

Objectives
- Identify the parts of a wave.
- Explain how the parts of a wave relate to wave movement.
- Describe how ocean waves form and move.
- Classify types of waves.

Terms to Learn
undertow
longshore current
whitecap
swell
tsunami
storm surge

READING STRATEGY

Prediction Guide Before reading this section, write the title of each heading in this section. Next, write what you think you will learn under each heading.

Waves

Have you ever seen a surfer riding waves? Did you ever wonder where the waves come from? And why are some waves big, while others are small?

We all know what ocean waves look like. Even if you've never been to the seashore, you've most likely seen waves on TV. But how do waves form and move? Waves are affected by a number of different factors. They can be formed by something as simple as wind or by something as violent as an earthquake. Ocean waves can travel through water slowly or incredibly quickly. Read on to discover the many forces that affect the formation and movement of ocean waves.

Anatomy of a Wave

Waves are made up of two main parts—crests and troughs. A *crest* is the highest point of a wave. A *trough* is the lowest point of a wave. Imagine a roller coaster designed with many rises and dips. The top of a rise on a roller-coaster track is similar to the crest of a wave, and the bottom of a dip in the track resembles the trough of a wave. The distance between two adjacent wave crests or wave troughs is a *wavelength*. The vertical distance between the crest and trough of a wave is called the *wave height*. **Figure 1** shows the parts of a wave.

✓ **Reading Check** What is the lowest point of a wave called? (*See the Appendix for answers to Reading Checks.*)

| Figure 1 | Parts of a Wave |

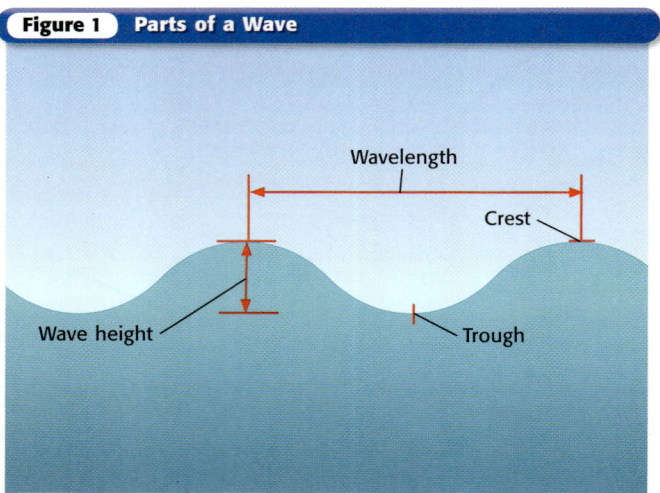

Wavelength

Crest

Wave height

Trough

CHAPTER RESOURCES

Chapter Resource File
- **Lesson Plan**
- **Directed Reading A** BASIC
- **Directed Reading B** SPECIAL NEEDS

Technology

Transparencies
- Bellringer
- Determining Wave Period

Answer to Reading Check

The lowest point of a wave is called a *trough*.

Wave Formation and Movement

If you have watched ocean waves before, you may have noticed that water appears to move across the ocean's surface. However, this movement is only an illusion. Most waves form as wind blows across the water's surface and transfers energy to the water. As the energy moves through the water, so do the waves. But the water itself stays behind, rising and falling in circular movements. Notice in **Figure 2** that the floating bottle remains in the same spot as the waves travel from left to right. This circular motion gets smaller as the water depth increases, because wave energy decreases as the water depth increases. Wave energy reaches only a certain depth. Below that depth, the water is not affected by wave energy.

Specifics of Wave Movement

Waves not only come in different sizes but also travel at different speeds. To calculate wave speed, scientists must know the wavelength and the wave period. *Wave period* is the time between the passage of two wave crests (or troughs) at a fixed point, as shown in **Figure 3**. Dividing wavelength by wave period gives you wave speed, as shown below.

$$\frac{\text{wavelength (m)}}{\text{wave period (s)}} = \text{wave speed (m/s)}$$

For any given wavelength, an increase in the wave period will decrease the wave speed and a decrease in the wave period will increase the wave speed.

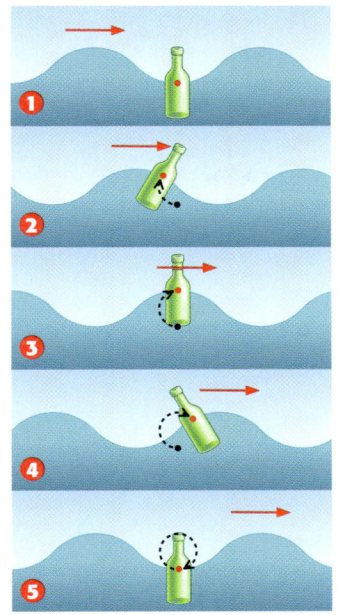

Figure 2 *Like the bottle in this figure, water remains in the same place as waves travel through it.*

Figure 3　Determining Wave Period

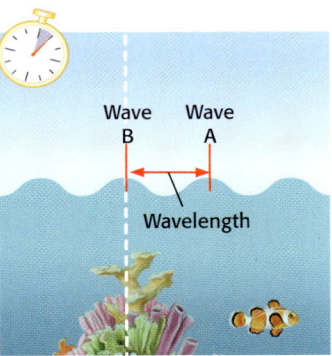

❶ Notice that the waves are moving from left to right.

❷ The clock begins running as Wave A passes the reef's peak.

❸ The clock stops as Wave B passes the reef's peak. The time shown on the clock (5 s) represents the wave period.

Types of Waves Have students reproduce **Figures 4** and **5.** As you read the section, have students label their diagrams. Use the teaching transparency entitled "Measuring Wavelength; Measuring Frequency" to review these concepts.

🄛🅂 **Visual/Intrapersonal** | English Language Learners

Offshore Breakers Ask the class why sighting a line of offshore breakers might cause sailors to consider turning their boats around. (As water becomes shallower, the wave height increases and the waves may break. Breaking waves could indicate a submerged sandbar or reef.)

CONNECTION ACTIVITY
Real World —— GENERAL

Water Safety Discuss with students what they should do if they are ever caught in a rip current. Explain that instead of trying to swim against the current, they should signal for help and swim parallel to the shore. This will get them out of the rip current. They can then swim to shore with the waves. Diagram this scenario for students, and have them show you the proper direction to swim. 🄛🅂 **Verbal**

Doing the Wave

1. Tie one end of a **thin piece of rope** to a **doorknob**.
2. Tie a **ribbon** around the rope halfway between the doorknob and the other end of the rope.
3. Holding the rope at the untied end, quickly move the rope up and down and observe the ribbon.
4. How does the movement of the rope and ribbon relate to the movement of water and deep-water waves?
5. Repeat step 3, but move the rope higher and lower this time.
6. How does this affect the waves in the rope?

Types of Waves

As you learned earlier in this section, wind forms most ocean waves. Waves can also form by other mechanisms. Underwater earthquakes and landslides as well as impacts by cosmic bodies can form different types of waves. Most waves move in one way regardless of how they are formed. Depending on their size and the angle at which they hit the shore, waves can generate a variety of near-shore events, some of which can be dangerous to humans.

Deep-Water Waves and Shallow-Water Waves

Have you ever wondered why waves increase in height as they approach the shore? The answer has to do with the depth of the water. *Deep-water waves* are waves that move in water deeper than one-half their wavelength. When the waves reach water shallower than one-half their wavelength, they begin to interact with the ocean floor. These waves are called *shallow-water waves*. **Figure 4** shows how deep-water waves become shallow-water waves as they move toward the shore.

As deep-water waves become shallow-water waves, the water particles slow down and build up. This change forces more water between wave crests and increases wave height. Gravity eventually pulls the high wave crests down, which causes them to crash into the ocean floor as *breakers*. The area where waves first begin to tumble downward, or break, is called the *breaker zone*. Waves continue to break as they move from the breaker zone to the shore. The area between the breaker zone and the shore is called the *surf*.

✓ **Reading Check** How do deep-water waves become shallow-water waves?

Figure 4 **Deep-Water and Shallow-Water Waves**

Deep-water waves become shallow-water waves when they reach depths of less than half of their wavelength.

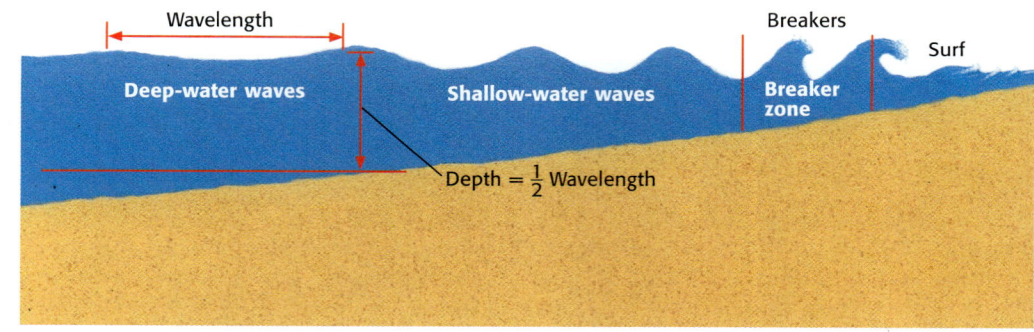

Answer to Reading Check

Deep-water waves become shallow-water waves as they move toward the shore and reach water that is shallower than one-half their wavelength.

Answer to Quick Lab

4. Sample answer: As wave energy passed through the rope, the ribbon moved up and down, but it did not move closer to the doorknob. Similarly, as wave energy passes through water, it moves it up and down but does not transport water.
6. Sample answer: The wave height increased.

Figure 5 Formation of an Undertow

Head-on waves create an undertow.

Direction of wave movement

Undertow

Shore Currents

When waves crash on the beach head-on, the water they moved through flows back to the ocean underneath new incoming waves. This movement of water, which carries sand, rock particles, and plankton away from the shore, is called an **undertow**. **Figure 5** illustrates the back-and-forth movement of water at the shore.

Longshore Currents

When waves hit the shore at an angle, they cause water to move along the shore in a current called a **longshore current**, which is shown in **Figure 6.** Longshore currents transport most of the sediment in beach environments. This movement of sand and other sediment both tears down and builds up the coastline. Unfortunately, longshore currents also carry and spread trash and other types of ocean pollution along the shore.

undertow a subsurface current that is near shore and that pulls objects out to sea

longshore current a water current that travels near and parallel to the shoreline

Longshore current

Figure 6 Longshore currents form where waves approach beaches at an angle.

Close

Reteaching — BASIC

Wave Components Ask students to describe in their own words the four components of a wave. LS Verbal

Quiz — GENERAL

Have students create a concept map linking section concepts and vocabulary.

Alternative Assessment — GENERAL

Diagraming Waves Have students use the wave characteristics described in this section to construct diagrams of different types of waves with various characteristics. The diagrams should be made on poster board so that they can be shared with the class and displayed in the classroom. LS Visual

Homework — GENERAL

Writing **Tsunamis** Have students research a tsunami that occurred and write a brief newspaper-style article about it. The articles should explain the cause and effects of the tsunami. Students could learn about the tsunami that struck the Pacific coast of Nicaragua in 1992 or the one that devastated Papua New Guinea in 1998. Encourage students to read their articles to the class. LS Verbal/Intrapersonal

Figure 7 Whitecaps (left) break in the open ocean, while swells, (right), roll gently in the open ocean.

whitecap the bubbles in the crest of a breaking wave

swell one of a group of long ocean waves that have steadily traveled a great distance from their point of generation

tsunami a giant ocean wave that forms after a volcanic eruption, submarine earthquake, or landslide

Open-Ocean Waves

Sometimes waves called *whitecaps* form in the open ocean. **Whitecaps** are white, foaming waves with very steep crests that break in the open ocean before the waves get close to the shore. These waves usually form during stormy weather, and they are usually short-lived. Winds that are far away from the shore form waves called *swells*. **Swells** are rolling waves that move steadily across the ocean. Swells have longer wavelengths than whitecaps and can travel for thousands of kilometers. **Figure 7** shows how whitecaps and swells differ.

Tsunamis

Professional surfers often travel to Hawaii to catch some of the highest waves in the world. But even the best surfers would not be able to handle a tsunami. **Tsunamis** are waves that form when a large volume of ocean water is suddenly moved up or down. This movement can be caused by underwater earthquakes, volcanic eruptions, landslides, underwater explosions, or the impact of a meteorite or comet. The majority of tsunamis occur in the Pacific Ocean because of the large number of earthquakes in that region. **Figure 8** shows how an earthquake can generate a tsunami.

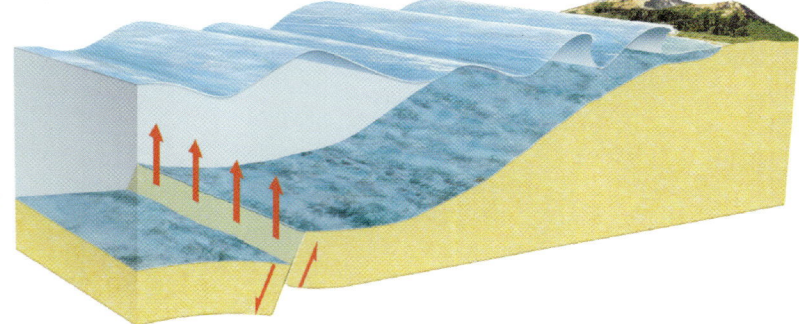

Figure 8 An upward shift in the ocean floor creates an earthquake. The energy released by the earthquake pushes a large volume of water upward, which creates a series of tsunamis.

Answer to Reading Check

A storm surge is a local rise in sea level near the shore and is caused by strong winds from a storm, such as a hurricane. Storm surges are difficult to study because they disappear as quickly as they form.

CONNECTION to Meteorology — ADVANCED

NOAA The National Oceanic and Atmospheric Administration (NOAA) studies oceanography and meteorology. Founded in 1970, this federal agency forecasts weather and monitors potentially destructive natural events, such as hurricanes, floods, and tsunamis. Interested students can write to NOAA or visit its Web site for more information.

Storm Surges

A local rise in sea level near the shore that is caused by strong winds from a storm, such as a hurricane, is called a **storm surge.** Winds form a storm surge by blowing water into a big pile under the storm. As the storm moves onto shore, so does the giant mass of water beneath it. Storm surges often disappear as quickly as they form, which makes them difficult to study. Storm surges contain a lot of energy and can reach about 8 m in height. Their size and power often make them the most destructive part of hurricanes.

storm surge a local rise in sea level near the shore that is caused by strong winds from a storm, such as those from a hurricane

✔ **Reading Check** What is a storm surge? Why are storm surges difficult to study?

SECTION Review

Summary

- Waves are made up of two main parts—crests and troughs.
- Waves are usually created by the transfer of the wind's energy across the surface of the ocean.
- Waves travel through water near the water's surface, while the water itself rises and falls in circular movements.
- Wind-generated waves are classified as deep-water or shallow-water waves.
- When waves hit the shore at a certain angle, they can create either an undertow or a longshore current.
- Tsunamis are dangerous waves that can be very destructive to coastal communities.

Using Key Terms

For each pair of terms, explain how the meanings of the terms differ.

1. *whitecap* and *swell*
2. *undertow* and *longshore current*
3. *tsunami* and *storm surge*

Understanding Key Ideas

4. Longshore currents transport sediment
   a. to the open ocean.
   b. along the shore.
   c. only during low tide.
   d. only during high tide.
5. Where do deep-water waves become shallow-water waves?
6. Explain how water moves as waves travel through it.
7. Name five events that can cause a tsunami.
8. Describe the two parts of a wave.

Math Skills

9. If a barrier island that is 1 km wide and 10 km long loses 1.5 m of its width per year to erosion by a longshore current, how long will the island take to lose one-fourth of its width?

Critical Thinking

10. **Analyzing Processes** How would you explain a bottle moving across the water in the same direction that the waves are traveling? Make a drawing of the bottle's movement.
11. **Analyzing Processes** Describe the motion of a wave as it approaches the shore.
12. **Applying Concepts** Explain how energy plays a role in the creation of ocean waves.
13. **Making Comparisons** How does the formation of an undertow differ from the formation of a longshore current? How is sand on the beach affected by each?

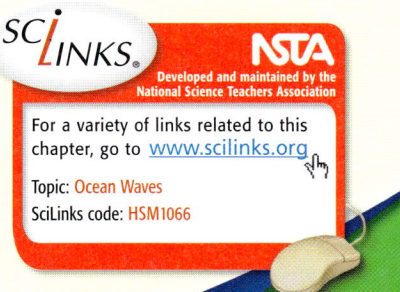

For a variety of links related to this chapter, go to www.scilinks.org

Topic: Ocean Waves
SciLinks code: HSM1066

Answers to Section Review

1. Sample answer: A whitecap is a wave that is white and foamy and has very steep crests that break in the open ocean. Swells are rolling waves that move steadily across the ocean.
2. Sample answer: An undertow is a movement of water that carries sand, rock particles, and plankton away from the shoreline. A longshore current is a water current that travels near and parallel to the shoreline.
3. Sample answer: A tsunami is a giant ocean wave that forms after a volcanic eruption, submarine earthquake, or landslide. A storm surge is a local rise in sea level near the shore and is caused by a storm, such as a hurricane.
4. b
5. Deep-water waves become shallow-water waves as they move toward the shore.
6. Water at and near the surface rises and falls in circular movements as waves move through it.
7. underwater earthquakes, volcanic eruptions, landslides, underwater explosions, and the impact of a meteorite or comet
8. The crest is the highest point of a wave, and the trough is the lowest point.
9. $1/4 \times 1 \text{ km} = 0.25 \text{ km}$;
   $1.5 \text{ m/year} \div 1000 \text{ m/km} = 0.0015 \text{ km/year}$;
   $0.25 \text{ km} \div 0.0015 \text{ km/year} =$ about 167 years
10. A floating bottle remains in the same spot as waves travel from left to right. If the bottle moves in the same direction as the waves, it is moving because of a surface current. Student drawings should look similar to **Figure 2** but should include a surface current.
11. As a wave approaches the shore, wave height increases and causes the wave to crash into the ocean floor. Wave height decreases and the wave continues to break until it reaches the shore.
12. Waves are created when energy is transferred to the water as wind blows across the water's surface.
13. An undertow forms when waves crash on the beach head-on. A longshore current forms when waves hit the shore at an angle. An undertow carries sand away from the shoreline, and a longshore current carries sand along the shoreline.

SECTION
4

Focus

Overview

Students will learn how the position of Earth in relation to the moon and the sun creates different kinds of tides. They will also learn about the effects of coastal topography on tidewaters.

🔔 Bellringer

Show students a golf ball. Tell them that if the moon had the mass of a golf ball, the sun would have the mass of approximately 110 school buses. Ask students why they think that the moon exerts more influence over tides on Earth than the sun does. Tell students to write their guesses on a sheet of paper. (Gravitational force decreases with distance. The sun is almost 150 million kilometers from the Earth, but the moon is only 385,000 km away.)

Motivate

ACTIVITY ——————— GENERAL

Intertidal Zones If possible, plan a visit to a coastal area. Beforehand, have the class research intertidal zones and the organisms that inhabit these areas. If a visit to the coast is not possible, have students research intertidal organisms and their survival strategies. Students should share their findings with the class. **LS** Interpersonal

READING WARM-UP

Objectives

- Explain tides and their relationship with the Earth, sun, and moon.
- Describe four different types of tides.
- Analyze the relationship between tides and coastal land.

Terms to Learn

tide   spring tide
tidal range   neap tide

READING STRATEGY

Discussion Read this section silently. Write down questions that you have about this section. Discuss your questions in a small group.

tide the periodic rise and fall of the water level in the oceans and other large bodies of water

Tides

If you stand at some ocean shores long enough, you will see the edge of the ocean shrink away from you. Wait longer, and you will see it return to its original place on the shore. Would you believe the moon causes this movement?

You have learned how winds and earthquakes can move ocean water. But less obvious forces move ocean water in regular patterns called tides. **Tides** are daily changes in the level of ocean water. Tides are influenced by the sun and the moon, as shown in **Figure 1,** and they occur in a variety of cycles.

The Lure of the Moon

The phases of the moon and their relationship to the tides were first discovered more than 2,000 years ago by a Greek explorer named *Pytheas.* But Pytheas and other early investigators could not explain the relationship. A scientific explanation was not given until 1687, when Sir Isaac Newton's theories on the principle of gravitation were published.

The gravity of the moon pulls on every particle of the Earth. But the pull on liquids is much more noticeable than on solids, because liquids move more easily. Even the liquid in an open soft drink is slightly pulled by the moon's gravity.

✓ **Reading Check** How does the moon affect Earth's particles? (*See the Appendix for answers to Reading Checks.*)

High Tide and Low Tide

How often tides occur and the difference in tidal levels depend on the position of the moon as it revolves around the Earth. The moon's pull is strongest on the part of the Earth directly facing the moon.

Figure 1 *Although gravitational forces from both the sun and moon continuously pull on the Earth, the moon's gravity is the dominant force on Earth's tides.*

CHAPTER RESOURCES

Chapter Resource File

- **Lesson Plan**
- **Directed Reading A** BASIC
- **Directed Reading B** SPECIAL NEEDS

Technology

Transparencies
- Bellringer

Answer to Reading Check

The gravity of the moon pulls on every particle of the Earth.

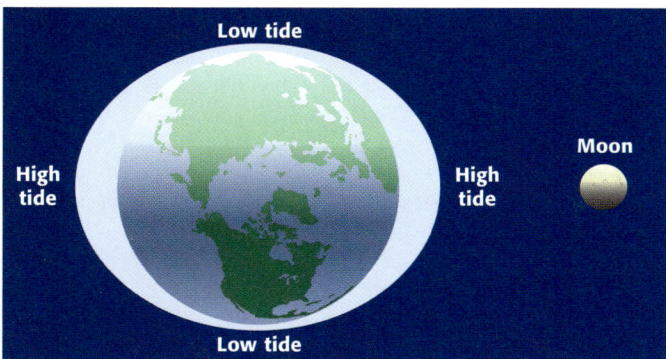

Figure 2 *High tide occurs on the part of Earth that is closest to the moon. At the same time, high tide also occurs on the opposite side of Earth.*

Battle of the Bulge

When part of the ocean is directly facing the moon, the water there bulges toward the moon. At the same time, water on the opposite side of the Earth bulges because of the rotation of the Earth and the motion of the moon around the Earth. These bulges are called *high tides*. Notice in **Figure 2** how the position of the moon causes the water to bulge. Also notice that when high tides occur, water is drawn away from the area between the high tides, which causes *low tides* to form.

Timing the Tides

The rotation of the Earth and the moon's revolution around the Earth determine when tides occur. If the Earth rotated at the same speed that the moon revolves around the Earth, the tides would not alternate between high and low. But the moon revolves around the Earth much more slowly than the Earth rotates. As **Figure 3** shows, a spot on Earth that is facing the moon takes 24 h and 50 m to rotate and face the moon again.

Figure 3 *Tides occur at different locations on Earth because the Earth rotates more quickly than the moon revolves around the Earth.*

Tuesday, 11:00 AM

Wednesday, 11:50 AM

CONNECTION TO Language Arts

WRITING SKILL **Mont-St-Michel Is Sometimes an Island?** Mont-St-Michel is located off the coast of France. Mont-St-Michel experiences extreme tides. The tides are so extreme that during high tide, it is an island and during low tide, it is connected to the mainland. Research the history behind Mont-St-Michel and then write a short story describing what it would be like to live there for a day. Be sure to include a description of Mont-St-Michel at high tide and at low tide.

Teach

Using the Figure — GENERAL

Earth's Rotation Remind students that Earth rotates much faster than the moon orbits. In **Figure 3,** point out that the distance that the moon traveled is much shorter than the distance the spot on Earth that faces the moon traveled. The spot on Earth made more than a full rotation around Earth's axis.
LS Visual

Discussion — BASIC

Tide Discussion Help students understand the hypothetical scenario discussed in the text (in which the moon orbits Earth at the same speed that Earth rotates). Have students imagine that the moon is attached to a pole that is anchored to Earth. In this scenario, the moon would turn with Earth as Earth rotates. The moon would always be facing the same spot on Earth, so high tide would always occur at that spot and at the spot on the opposite side of Earth. In other words, the moon would orbit the Earth every 24 hours. Make sure students realize that this is not the case and that the location of tides varies constantly because the moon's revolution and the Earth's rotation occur at different speeds. **LS Verbal**

SCIENTISTS AT ODDS

Galileo Versus Kepler The cause of tides was a debated topic in the 16th and 17th centuries. Galileo suggested that there was a connection between tides and Earth's motion. To Galileo, tides proved beyond all doubt that Earth was moving. He argued that because Earth's waters are moving, Earth must be moving too. Johannes Kepler, another prominent scientist of the time, argued that the tides were linked to the moon's phases. Galileo made such a convincing argument that Kepler's ideas were dismissed. It was not until after Newton that scientists accepted that the gravitational forces exerted by the moon and the sun, as well as the rotation of Earth, are responsible for the tides.

Close

Reteaching — BASIC

Spring Tides and Neap Tides
Have students reproduce **Figure 4** in their **science journal.** Then, have students write a definition for *spring tides* and *neap tides* in their own words. **English Language Learners**
LS Visual

Quiz — GENERAL

1. Why do spring tides exhibit such extremes of range? (because the gravitational force is greater than when neap tides occur)

2. What is a tidal bore? (a body of water that rushes up through a narrow bay, estuary, or river)

Alternative Assessment — GENERAL

Illustrating Tidal Relationships
Draw a random configuration of the sun, the moon, and Earth. Have students copy the drawing onto a sheet of paper. Then, ask them to draw the tidal bulges caused by that configuration. Finally, have students draw the positions of the sun, the moon, and Earth during spring and neap tides. **LS Visual/Verbal**

---

tidal range the difference in levels of ocean water at high tide and low tide

spring tide a tide of increased range that occurs two times a month, at the new and full moons

neap tide a tide of minimum range that occurs during the first and third quarters of the moon

---

Tidal Variations

The sun also affects tides. The sun is much larger than the moon, but the sun is also much farther away. As a result, the sun's influence on tides is less powerful than the moon's influence. The combined forces of the sun and the moon on the Earth result in tidal ranges that vary based on the positions of all three bodies. A **tidal range** is the difference between levels of ocean water at high tide and low tide.

✓ Reading Check What is a tidal range?

Spring Tides

When the sun, Earth, and moon are aligned, spring tides occur. **Spring tides** are tides with the largest daily tidal range and occur during the new and full moons, or every 14 days. The first time spring tides occur is when the moon is between the sun and Earth. The second time spring tides occur is when the moon and the sun are on opposite sides of the Earth. **Figure 4** shows the positions of the sun, Earth, and moon during spring tides.

Neap Tides

When the sun, Earth, and moon form a 90° angle, neap tides occur. **Neap tides** are tides with the smallest daily tidal range and occur during the first and third quarters of the moon. Neap tides occur halfway between the occurrence of spring tides. When neap tides occur, the gravitational forces on the Earth by the sun and moon work against each other. **Figure 4** shows the positions of the sun, Earth, and moon during neap tides.

Figure 4 Spring Tides and Neap Tides

Spring Tides During spring tides, the gravitational forces of the sun and moon pull on the Earth either from the same direction (left) or from opposite directions (right).

Neap Tides During neap tides, the sun and moon are at right angles with respect to the Earth. This arrangement lessens their gravitational effect on the Earth.

 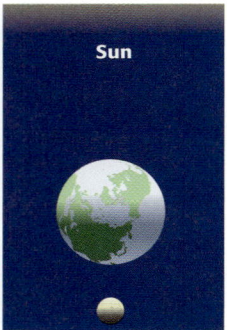  

Answer to Reading Check
A tidal range is the difference between levels of ocean water at high tide and low tide.

Homework — GENERAL

Graphing Help students look in the newspaper or on the Internet to find daily tidal information for a certain area for 1 month. Record the high- and low-tide measurements on a large chart that can be displayed in the classroom. At the end of the month, determine when spring tide and neap tide occurred. Compare these dates with the full-, new-, and quarter-moon dates on a calendar, and explain why the dates correspond. **LS Visual**

Tides and Topography

After a tidal range has been measured, the times that tides occur can be accurately predicted. This information can be useful for people who live near or visit the coast, as shown in **Figure 5**. In some coastal areas that have narrow inlets, movements of water called tidal bores occur. A *tidal bore* is a body of water that rushes up through a narrow bay, estuary, or river channel during the rise of high tide and causes a very sudden tidal rise. Tidal bores occur in coastal areas of China, the British Isles, France, and Canada.

Figure 5 *It's a good thing the people on this beach (left) knew when high tide occurred (right). These photos show the Bay of Fundy, in New Brunswick, Canada. The Bay of Fundy has the greatest tidal range on Earth.*

SECTION Review

Summary

- Tides are caused by the gravitational forces of the moon and sun on the Earth.
- The moon's gravity is the main force behind the tides.
- The positions of the sun and moon relative to the position of the Earth cause tidal ranges.
- The four different types of tides are: high tides, low tides, spring tides, and neap tides.

Using Key Terms

1. In your own words, write a definition for each of the following terms: *spring tides* and *neap tides*.

Understanding Key Ideas

2. Tides are at their highest during
   a. spring tide.
   b. neap tide.
   c. a tidal bore.
   d. the daytime.

3. Which tides have minimum tidal range? Which tides have maximum tidal range?

4. What causes tidal ranges?

Math Skills

5. If it takes 24 h and 50 min for a spot on Earth that is facing the moon to rotate to face the moon again, how many minutes does it take?

Critical Thinking

6. **Applying Concepts** How many days pass between the minimum and the maximum of the tidal range in any given area? Explain your answer.

7. **Analyzing Processes** Explain how the position of the moon relates to the occurrence of high tides and low tides.

SCI**LINKS**

NSTA
Developed and maintained by the National Science Teachers Association

For a variety of links related to this chapter, go to www.scilinks.org

Topic: Tides
SciLinks code: HSM1525

WEIRD SCIENCE

The moon also creates tides in our atmosphere that are called *lunar winds*. Lunar winds move eastward in the morning and westward in the evening. Although these tides travel only 0.08 km/h, they can be detected by studying slight fluctuations in weather patterns.

CHAPTER RESOURCES

Chapter Resource File
- Section Quiz **GENERAL**
- Section Review **GENERAL**
- Vocabulary and Section Summary **GENERAL**
- Critical Thinking **ADVANCED**

Technology
Transparencies
- Spring Tides and Neap Tides

Up from the Depths

Teacher's Notes

Time Required
One 45-minute class period

Lab Ratings

EASY ——————————— HARD

Teacher Prep 🧪
Student Set-Up 🧪🧪
Concept Level 🧪🧪
Clean Up 🧪🧪

MATERIALS

The materials listed on the student page are enough for a group of 4 to 5 students. Note that the food coloring is used only to distinguish the water layers. Any two colors will work. You may find it simpler to make a roll of plastic wrap available to the class. Groups can then take a piece when they reach the appropriate steps.

Safety Caution

Remind students to review all safety cautions and icons before beginning this lab activity.

OBJECTIVES

Demonstrate the effects of temperature and salinity on the density of water.

Describe why some parts of the ocean turn over, while others do not.

MATERIALS

- beakers, 400 mL (5)
- blue and red food coloring
- bucket of ice
- gloves, heat-resistant
- hot plate
- plastic wrap, 4 pieces, approximately 30 cm × 20 cm
- salt
- spoon
- tap water
- watch or clock

SAFETY

Up from the Depths

Every year, the water in certain parts of the ocean "turns over." That is, the water at the bottom rises to the top and the water at the top falls to the bottom. This yearly change brings fresh nutrients from the bottom of the ocean to the fish living near the surface. However, the water in some parts of the ocean never turns over. By completing this activity, you will find out why not.

Keep in mind that some parts of the ocean are warmer at the bottom, and some are warmer at the top. And sometimes the saltiest water is at the bottom and sometimes not. As you complete this activity, you will investigate how these factors help determine whether the water will turn over.

Ask a Question

1. Why do some parts of the ocean turn over and not others?

Form a Hypothesis

2. Write a hypothesis that is a possible answer to the question above. Explain your reasoning.

Test the Hypothesis

3. Label the beakers 1 through 5. Fill beakers 1 through 4 with tap water.

4. Add a drop of blue food coloring to the water in beakers 1 and 2, and stir with the spoon.

5. Place beaker 1 in the bucket of ice for 10 min.

6. Add a drop of red food coloring to the water in beakers 3 and 4, and stir with the spoon.

7. Set beaker 3 on a hot plate turned to a low setting for 10 min.

8. Add one spoonful of salt to the water in beaker 4, and stir with the spoon.

Gordon Zibelman
Drexel Hill Middle School
Drexel Hill, Pennsylvania

CHAPTER RESOURCES

Chapter Resource File

 • Datasheet for Chapter Lab
• Lab Notes and Answers

Technology

 Classroom Videos
• Lab Videos for Earth Science

• Turning the Tides

9 While beaker 1 is cooling and beaker 3 is heating, copy the observations table below on a sheet of paper.

Observations Table	
Mixture of water	Observations
Warm water placed above cold water	
Cold water placed above warm water	*DO NOT WRITE IN BOOK*
Salty water placed above fresh water	
Fresh water placed above salty water	

10 Pour half of the water in beaker 1 into beaker 5. Return beaker 1 to the bucket of ice.

11 Tuck a sheet of plastic wrap into beaker 5 so that the plastic rests on the surface of the water and lines the upper half of the beaker.

12 Put on your gloves. Slowly pour half of the water in beaker 3 into the plastic-lined upper half of beaker 5 to form two layers of water. Return beaker 3 to the hot plate, and remove your gloves.

13 Very carefully, pull on one edge of the plastic wrap and remove it so that the warm, red water rests on the cold, blue water.
Caution: The plastic wrap may be warm.

14 Wait about 5 minutes, and then observe the layers in beaker 5. Did one layer remain on top of the other? Was there any mixing or turning over? Record your observations in your observations table.

15 Empty beaker 5, and rinse it with clean tap water.

16 Repeat the procedure used in steps 10–15. This time, pour warm, red water from beaker 3 on the bottom and cold, blue water from beaker 1 on top. (Use gloves when pouring warm water.)

17 Again, repeat the procedure used in steps 10–15. This time, pour blue tap water from beaker 2 on the bottom and red, salty water from beaker 4 on top.

18 Repeat the procedure used in steps 10–15 a third time. This time, pour red, salty water from beaker 4 on the bottom and blue tap water from beaker 2 on top.

Analyze the Results

1 **Analyzing Data** Compare the results of all four trials. Explain why the water turned over in some of the trials but not in all of them.

Draw Conclusions

2 **Evaluating Results** What is the effect of temperature and salinity on the density of water?

3 **Drawing Conclusions** What makes the temperature of ocean water decrease? What could make the salinity of ocean water increase?

4 **Drawing Conclusions** What reasons can you give to explain why some parts of the ocean do not turn over in the spring while some do?

Applying Your Data
Suggest a method for setting up a model that tests the combined effects of temperature and salinity on the density of water. Consider using more than two water samples and dyes.

Procedure

14. The warm, red water remained on top of the cold, blue water. There was very little mixing (if any) and no turning over.

16. The cold, blue water did not remain on top of the warm, red water. There was very little mixing (if any), and the water turned over.

17. The red, salty water did not remain on top of the blue tap water. There was little mixing, and the water turned over.

18. The blue tap water remained on top of the red, salty water. There was little mixing, and the water did not turn over.

Analyze the Results

1. In each case, the denser water sank to the bottom. Cold water is denser than warm water. When put in a beaker with warm water, the cold water either stayed at the bottom or sank to the bottom. Salt water is denser than fresh water. When put in a beaker with fresh water, the salt water either stayed at the bottom or sank to the bottom.

Draw Conclusions

2. The density of water increases as the water's temperature decreases—cold water is denser than warm water. The density of water increases as the water's salinity increases—salt water is denser than fresh water.

3. The temperature of water can decrease because of seasonal temperature fluctuations or cold wind blowing across the water's surface. Currents can also carry cooler water to an area. The salinity of water can increase when evaporation occurs or when ice forms on the water's surface. These processes leave salts behind and make the remaining water denser.

4. Parts of the ocean that turn over do so because their density changes because of variations in salinity or temperature.

Applying Your Data
The following combinations could be used:

Top	Bottom
salt/cold	fresh/warm
salt/warm	fresh/cold
fresh/cold	salt/warm
fresh/warm	salt/cold

Chapter Review

Assignment Guide

SECTION	QUESTIONS
1	1, 5, 15, 16, 18, 21–23
2	2, 10, 11, 19
3	6, 7, 12, 14
4	3, 4, 8, 9, 13, 20
1, 3, 4	17

ANSWERS

Using Key Terms

1. Sample answer: Surface currents are streamlike movements of water that occur at or near the surface of the ocean. Deep currents are streamlike movements of ocean water far below the surface.

2. Sample answer: El Niño is a change in the water temperature in the Pacific Ocean that produces a warm current. La Niña is a change in the eastern Pacific Ocean in which the surface water temperature becomes unusually cool.

3. Sample answer: A spring tide is a tide of increased range that occurs during the full and new moons. A neap tide is a tide of minimum tidal range that occurs during the first and third quarters of the moon.

4. Sample answer: A tide is the periodic rise and fall of the water level in the oceans and other large bodies of water. Tidal range is the difference in levels of ocean water at high tide and low tide.

USING KEY TERMS

For each pair of terms, explain how the meanings of the terms differ.

1 *surface current* and *deep current*

2 *El Niño* and *La Niña*

3 *spring tide* and *neap tide*

4 *tide* and *tidal range*

UNDERSTANDING KEY IDEAS

Multiple Choice

5 Deep currents form when
  a. cold air decreases water density.
  b. warm air increases water density.
  c. the ocean surface freezes and solids from the water underneath are removed.
  d. salinity increases.

6 When waves come near the shore,
  a. they speed up.
  b. they maintain their speed.
  c. their wavelength increases.
  d. their wave height increases.

7 Whitecaps break
  a. in the surf.
  b. in the breaker zone.
  c. in the open ocean.
  d. as their wavelength increases.

8 Tidal range is greatest during
  a. spring tide.
  b. neap tide.
  c. a tidal bore.
  d. the daytime.

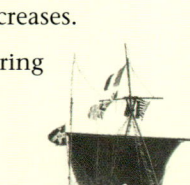

9 Tides alternate between high and low because the moon revolves around the Earth
  a. at the same speed the Earth rotates.
  b. at a much faster speed than the Earth rotates.
  c. at a much slower speed than the Earth rotates.
  d. at different speeds.

10 El Niño can cause
  a. droughts to occur in Indonesia and Australia.
  b. upwelling to occur off the coast of South America.
  c. earthquakes.
  d. droughts to occur in the southern half of the United States.

Short Answer

11 Explain the relationship between upwelling and El Niño.

12 Describe the two parts of a wave. Describe how these two parts relate to wavelength and wave height.

13 Compare the relative positions of the Earth, moon, and sun during the spring and neap tides.

14 Explain the difference between the breaker zone and the surf.

15 Describe how warm-water currents affect the climate in the British Isles.

16 Describe the factors that form deep currents.

Understanding Key Ideas

5. d

6. d

7. c

8. a

9. c

10. a

11. When El Niño occurs, warm surface water remains along the Pacific coast of South America. Therefore, upwelling does not occur along the coast.

12. A crest is the highest point of a wave, and a trough is the lowest point of a wave. The wavelength of a wave is the distance between two adjacent wave crests or wave troughs. Wave period is the time between the passage of two wave crests or troughs. Wavelength divided by wave period yields wave speed.

13. During neap tide, the sun, the moon, and Earth form a right angle, with Earth in the middle. During spring tide, the sun, moon, and Earth align in a straight line.

CRITICAL THINKING

17 **Concept Mapping** Use the following terms to create a concept map: *wind, deep currents, sun's gravity, types of ocean-water movement, surface currents, tides, increasing water density, waves,* and *moon's gravity*.

18 **Identifying Relationships** Why are tides more noticeable in Earth's oceans than on its land?

19 **Expressing Opinions** Explain why it's important to study El Niño and La Niña.

20 **Applying Concepts** Suppose you and a friend are planning a fishing trip to the ocean. Your friend tells you that the fish bite more in his secret fishing spot during low tide. If low tide occurred at the spot at 7 a.m. today and you are going to fish there in 1 week, at what time will low tide occur in that spot?

21 **Identifying Relationships** Describe how global winds, the Coriolis Effect, and continental deflections form a pattern of surface currents on Earth.

INTERPRETING GRAPHICS

The diagram below shows some of Earth's major surface currents that flow in the Western Hemisphere. Use the diagram to answer the questions that follow.

Arctic Ocean

Oyashio
North America
Labrador
N. Atlantic Drift
North Pacific Ocean
Gulf Stream
North Atlantic Ocean
Canary
California
North Equatorial
North Equatorial
Equatorial Countercurrent
Equator
Equatorial Countercurrent
South America
South Equatorial
South Equatorial
Peru
South Pacific Ocean
South Atlantic Ocean

→ Warm current
→ Cold current

Antarctic Circumpolar

22 List two warm-water currents and two cold-water currents.

23 How do you think the Labrador Current affects the climate of Canada and Greenland?

Critical Thinking

17. 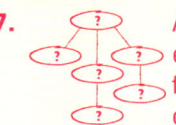 An answer to this exercise can be found at the end of this book.

18. Because the oceans are liquid, they flow more easily than land.

19. It is important to study El Niño and La Niña because both affect climate, land, and organisms.

20. 12:50 P.M. (The answer 1:15 A.M. is also acceptable.)

21. Global winds blow across Earth's surface, which creates surface currents in the ocean. The Coriolis effect is the apparent curving of moving objects from a straight path due to Earth's rotation. Therefore, the rotation of the Earth causes ocean currents to curve as they move across Earth's surface. Continental deflections occur when a surface current comes into contact with a continent. As a result, surface currents deflect, or change direction.

Interpreting Graphics

22. Answers may vary. Any two currents marked by red arrows are acceptable as warm-water currents. Any two currents marked by blue arrows are acceptable as cold-water currents.

23. Sample answer: Because it is a cold-water current, the Labrador Current would most likely bring a cool climate to Canada and Greenland.

14. The breaker zone is the zone where waves first begin to tumble downward. The surf is the zone between the breaker zone and the shore. In the surf, water moves toward the shore.

15. The Gulf Stream, which is a warm-water current, creates a relatively mild climate for the British Isles.

16. Deep currents form where the density of ocean water increases. Water density depends on temperature and salinity.

CHAPTER RESOURCES

Chapter Resource File
- Chapter Review GENERAL
- Chapter Test A GENERAL
- Chapter Test B ADVANCED
- Chapter Test C SPECIAL NEEDS
- Vocabulary Activity GENERAL

Workbooks

Study Guide
- Assessment resources are also available in Spanish.

Teacher's Note

To provide practice under more realistic testing conditions, give students 20 minutes to answer all of the questions in this Standardized Test Preparation.

Answers to the standardized test preparation can help you identify student misconceptions and misunderstandings.

READING

Passage 1

1. B
2. I

➕ TEST DOCTOR

Question 1: Although the term *red tides* was used to describe algal blooms, the abbreviation HABs is more accurate because algal blooms are not always red and are not directly related to tides. This fact is mentioned in the first paragraph.

Question 2: The fact is mentioned in the last paragraph: "Unfortunately, seafood contamination is not noticeable without testing . . ." Although HABs are reddish in color, contaminated seafood does not appear different physically.

Passage 2

1. D
2. G

READING

Read each of the passages below. Then, answer the questions that follow each passage.

Passage 1 When certain algae grow rapidly, they clump together on the ocean's surface in an algal bloom that changes the color of the water. Because these algal blooms often turn the water red or reddish brown and tidal conditions were believed to cause the blooms, people called these blooms *red tides*. However, algal blooms are not always red and are not directly related to tides. Scientists now call these algae clusters <u>harmful algal blooms (HABs)</u>. HABs are considered harmful because the species of algae that makes up the blooms produces toxins that can poison fish and shellfish, which in turn can poison people.

Unfortunately, seafood contamination is not noticeable without testing and is not easily eliminated. The toxins don't change the flavor of the seafood, and cooking the seafood doesn't eliminate the toxins.

1. Why did scientists start calling red tides *HABs*?
   A The name *HABs* is easier to remember.
   B The name *red tides* was not accurate in describing the phenomenon.
   C The algal blooms are actually green.
   D The term *red tides* did not reflect the danger of the blooms.

2. How can a person tell if seafood has been contaminated by HABs?
   F Contaminated seafood has a reddish color.
   G HABs change the flavor of the seafood.
   H Seafood contaminated by HABs has a strange smell.
   I Unfortunately, there is no easy way to tell.

Passage 2 Tsunamis are the most destructive waves in the ocean. Most tsunamis are caused by earthquakes on the ocean floor, but some can be caused by volcanic eruptions and underwater landslides. Tsunamis are sometimes called *tidal waves*, which is <u>misleading</u> because tsunamis have no connection with tides.

Tsunamis commonly have a wave period of about 15 min and a wave speed of about 725 km/h, which is about as fast as a jet airliner. By the time a tsunami reaches the shore, its height may be 30 to 40 m.

In 1960, a tsunami was triggered by an earthquake off the coast of South America. The tsunami was so powerful that it crossed the Pacific Ocean and hit the city of Hilo, on the coast of Hawaii, approximately 10,000 km away. The same tsunami then continued on to strike Japan.

1. The word *misleading* was used in this passage to describe the use of the term *tidal waves* because
   A tsunamis are related to tides.
   B tsunamis can cause extensive damage to shores.
   C tsunamis are related to earthquakes.
   D tsunamis are not related to tides.

2. Which of the following statements is a fact from the passage?
   F All tsunamis are caused by earthquakes.
   G A tsunami can travel as fast as a jet airliner.
   H The tsunami of 1960 caused destruction only in Japan.
   I Tsunamis are caused by surface currents.

➕ TEST DOCTOR

Question 2: This fact is mentioned in the second paragraph. Students may choose answer F because most tsunamis are caused by earthquakes. However, some tsunamis can be caused by events such as volcanic eruptions and underwater landslides. Answer H is incorrect because the last paragraph states that the tsunami of 1960 caused destruction in both Hawaii and Japan. Answer I is incorrect because surface currents are not related to tsunamis and are not mentioned in the passage.

The diagram below shows the possible positions of the moon relative to the Earth and sun during different tidal ranges. Use the diagram below to answer the questions that follow.

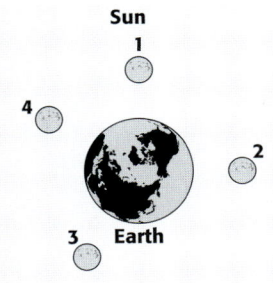

Sun
1

4

2

3 Earth

1. At which position would the moon be during a neap tide?

A 1

B 2

C 3

D 4

2. At which position would the moon be during a spring tide?

F 1

G 2

H 3

I 4

3. The tidal range would be greater when the moon is at position 3 than when the moon is at position 4 because

A position 4 forms a 90° angle with the sun and the Earth.

B position 3 is very near a neap-tide position.

C position 3 is very near a spring-tide position.

D position 4 is very near a spring-tide position.

Read each question below, and choose the best answer.

1. If a wave has a speed of 3 m/s and a wavelength of 12 m, what is its period? Use the following equation to answer the question above:

$$\frac{\text{wavelength (m)}}{\text{wave period (s)}} = \text{wave speed (m/s)}$$

A 36 s

B 4 m

C 24 s

D 4 s

2. Antarctic Bottom Water takes 750 years to move from the Antarctic coast to the equator. If the distance between the equator and the Antarctic coast is about 10,000 km, approximately how many kilometers does the bottom water move each year?

F 13 km

G 200 km

H 75 km

I 1 km

3. A boat is traveling north at 20 km/h against a current that is moving south at 12 km/h. What is the overall speed and direction of the boat?

A 8 km/h north

B 8 km/h south

C 32 km/h north

D 32 km/h south

4. Imagine that you are in a rowboat on the open ocean. You count 2 waves traveling right under your boat in 10 seconds. You estimate the wavelength to be 3 m. What is the wave speed?

F 0.6 m/s

G 6.0 m/s

H 0.3 m/s

I 3.0 m/s

Standardized Test Preparation

1. B

2. F

3. C

TEST DOCTOR

Question 3: Spring tides are tides with the largest daily tidal range. Spring tides occur when the sun and the moon pull on Earth from either the same direction or from opposite directions. Therefore, the moon at position 3 has the greatest tidal range—it is very near a spring-tide position because it is almost at a position opposite the sun.

1. D

2. F

3. A

4. F

TEST DOCTOR

Question 1: The first answer, 36 s, is incorrect because students would arrive at this answer by multiplying 3 by 12 rather than dividing 12 by 3. Answer B is incorrect because wave periods are measured in seconds, not in meters. Answer C is incorrect because it is too high. Answer D is correct because the wave period is 4 s; therefore, 12 m/ 4 s = 3 m/s.

Question 4: The first answer, F, is correct because students would arrive at this answer by first calculating the wave period by dividing 10 s by 2. Then, students would divide 3 m by 5 s to arrive at 0.6 m/s as the wave speed. Answer G is incorrect because the wave speed is too high. Answer H is incorrect because students would arrive at this answer by dividing 3 by 10 rather than dividing 3 by 5. Answer I is incorrect because the wave speed is too high.

Weird Science

ACTIVITY ————— GENERAL

Surface currents move in large, slow circles called *gyres*. In 1990, a Korean ship carrying a load of athletic shoes was bound for the United States. During a storm in the North Pacific Ocean, 80,000 shoes were washed overboard. Six months later, the shoes began to wash up on North American shores from Oregon to British Columbia. This accident turned out to be an advantage to oceanographers. They created a computer model to predict where the gyres would carry the shoes next. True to the model, the shoes began to wash up in Hawaii three years later. With the help of dedicated beach-combers, this accident has enabled oceanographers to create a detailed map of surface currents in the Pacific Ocean. Have students look at a map of global ocean currents and identify the currents that carried the shoes.

Answer to Math Activity
The average distance traveled by the toys per day was 10.7 km (3,220 km ÷ 301 days = 10.7 km).

Science in Action

Weird Science

Using Toy Ducks to Track Ocean Currents

Accidents can sometimes lead to scientific discovery. For example, on January 10, 1992, 29,000 plastic tub toys spilled overboard when a container ship traveling northwest of Hawaii ran into a storm. In November of that year, those toys began washing up on Alaskan beaches. When oceanographers heard about this, they placed advertisements in newspapers along the Alaskan coast asking people who found the toys to call them. Altogether, hundreds of toys were recovered. Using recovery dates and locations and computer models, oceanographers were able to re-create the toys' drift and figure out which currents carried the toys. As for the remaining toys, currents may carry them to a number of different destinations. Some may travel through the Arctic Ocean and eventually reach Europe!

Math ACTIVITY

Between January 10, 1992, and November 16, 1992, some of the toys were carried approximately 3,220 km from the cargo-spill site to the coast of Alaska. Calculate the average distance traveled by these toys per day. (Hint: The year 1992 was a leap year.)

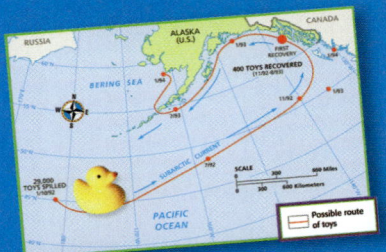

Science, Technology, and Society

Red Tides

Imagine going to the beach only to find that the ocean water has turned red and that a lot of fish are floating belly up. What could cause such damage to the ocean? It may surprise you to find that the answer is single-celled algae. When certain algae grow rapidly, they clump together on the ocean's surface in what are known as algal blooms. These algal blooms have been commonly called *red tides* because the blooms often turn the water red or reddish-brown. The term scientists use for these sudden explosions in algae growth is *harmful algal blooms* (HABs). The blooms are harmful because certain species of algae produce toxins that can poison fish, shellfish, and people who eat poisoned fish or shellfish. Toxic blooms can be carried hundreds of miles on ocean currents. HABs can ride into an area on an ocean current and cause fish to die and people who eat the poisoned fish or shellfish to become ill.

Social Studies ACTIVITY

Some scientists think that factors related to human activities, such as agricultural runoff into the ocean, are causing more HABs than occurred in the past. Other scientists disagree. Find out more about this issue, and have a class debate about the roles humans play in creating HABs.

Science, Technology, and Society

Background

Of the 4,400 phytoplankton species, a mere 50 to 60 can produce toxins. Even a small dose of the toxin of certain species can prove fatal. The best way for consumers to reduce the chances of buying poisoned seafood is to shop at a reputable seafood store. Seafood purchased at reputable supermarkets, restaurants, and seafood stores.

Answer to Social Studies Activity
At present, no definitive study links HABs to human activities; all information is still speculation. Activities considered to be either directly or indirectly related to the cause of HABs include overfishing, global warming, and ocean-nutrient fluctuations due to coastal development and water runoff.

Careers

Cristina Castro

Marine Biologist Have you ever imagined watching whales for a living? Cristina Castro does. Castro works as a marine biologist with the Pacific Whale Foundation in Ecuador. She is studying the migratory patterns of a whale species known as the *humpback whale*. Each year, the humpback whale migrates from feeding grounds in the Antarctic to the warm waters off Ecuador, where the whales breed. Her studies take place largely in the Machalilla National Park. The park is a two-mile stretch of beach that is protected by the government of Ecuador.

In her research, Cristina Castro focuses on the connection between El Niño events and the number of humpback whales in the waters off Ecuador. Castro believes that during an El Niño event, the waters off Ecuador are too hot for the whales. When the whales get hot, they have a difficult time cooling off because they have a thick coat of blubber that provides insulation. So, Castro believes that the whales stay in colder waters during an El Niño event.

Language Arts ACTIVITY

WRITING SKILL Research the humpback whale's migratory route from Antarctica to Ecuador. Write a short story in which you tell of the migration from the point of view of a young whale.

go.hrw.com
To learn more about these Science in Action topics, visit go.hrw.com and type in the keyword **HZ5H2OF**.

Current Science
Check out Current Science® articles related to this chapter by visiting go.hrw.com. Just type in the keyword **HZ5CS14**.

Answer to Language Arts Activity
Students can research information about the migration route of humpback whales by using the Internet or library resources. Encourage students to be creative when they write their short stories. Also, encourage students to read their short stories to the class.

Clean Up Your Act

Teacher's Notes

Time Required

Two 45-minute class periods

Lab Rating

EASY ━━━━━━━━━━━▶ HARD

Teacher Prep 🧪🧪🧪
Student Set-Up 🧪🧪🧪
Concept Level 🧪🧪🧪
Clean Up 🧪🧪🧪

MATERIALS

The materials listed on the student page are enough for a group of 4 to 5 students. To keep results consistent with all lab groups, each group should have the same-size gravel and the same-size sand. Also, the layers of sand and gravel should be the same dimensions in each group.

Safety Caution

Remind students to review all safety cautions and icons before beginning this lab activity. Caution students not to taste any of the liquids in this lab.

Clean Up Your Act

When you wash dishes, the family car, the bathroom sink, or your clothes, you wash them with water. But have you ever wondered how water gets clean? Two major methods of purifying water are filtration and evaporation. In this activity, you will use both of these methods to test how well they remove pollutants from water. You will test detritus (decaying plant matter), soil, vinegar, and detergent. Your teacher may also ask you to test other pollutants.

Form a Hypothesis

1 Form a hypothesis about whether filtration and evaporation will clean each of the four pollutants from the water and how well they might do it. Then, use the procedures below to test your hypothesis.

Part A: Filtration

Filtration is a common method of removing various pollutants from water. Filtration requires very little energy—gravity pulls water down through the layers of filter material. See how well this energy-efficient method works to clean your sample of polluted water.

Test the Hypothesis

2 Put on your gloves and goggles. Use scissors to carefully cut the bottom out of the empty soda bottle.

3 Using a small nail and hammer, carefully punch four or five small holes through the plastic cap of the bottle. Screw the plastic cap onto the bottle.

4 Turn the bottle upside down, and set its neck in a ring on a ring stand, as shown on the next page. Put a handful of gravel into the inverted bottle. Add a layer of activated charcoal, followed by thick layers of sand and gravel. Place a 400 mL beaker under the neck of the bottle.

5 Fill each of the large beakers with 1,000 mL of clean water. Set one beaker aside to serve as the control. Add three or four spoonfuls of each of the following pollutants to the other beaker: detritus, soil, household vinegar, and dishwashing detergent.

6 Copy the table on the next page, and record your observations for each beaker in the columns labeled "Before cleaning."

7 Observe the color of the water in each beaker.

8 Use a hand lens to examine the water for visible particles.

MATERIALS

Part A

- charcoal, activated
- goggles
- gravel
- hammer and small nail
- sand
- scissors
- soda bottle, plastic, with cap, 2 L

Part B

- bag, plastic sandwich, sealable
- flask, Erlenmeyer
- gloves, heat-resistant
- hot plate
- ice
- stopper, rubber, one-hole, with a glass tube
- tubing, plastic, 1.5 m

Parts A and B

- beaker, 400 mL
- beaker, 1,000 mL (2)
- detergent, dishwashing
- detritus (grass and leaf clippings)
- hand lens
- pH test strips
- ring stand with ring
- soil
- spoons, plastic (2)
- vinegar, household
- water, 2,000 mL

SAFETY

CLASSROOM TESTED & APPROVED

Kenneth Creese
White Mountain Jr. High
Rock Springs, Wyoming

9 Smell the water, and note any unusual odors.

10 Stir the water in each beaker rapidly with a plastic spoon, and check for suds. Use a different spoon for each sample.

11 Use a pH test strip to find the pH of the water.

12 Gently stir the clean water, and then pour half of it through the filtration device.

13 Observe the water in the collection beaker for color, particles, odors, suds, and pH. Be patient. It may take several minutes for the water to travel through the filtration device.

14 Record your observations in the appropriate "After filtration" column in your table.

15 Repeat steps 12–14 using the polluted water.

Analyze the Results

1 How did the color of the polluted water change after the filtration? Did the color of the clean water change?

2 Did the filtration method remove all of the particles from the polluted water? Explain.

3 How much did the pH of the polluted water change? Did the pH of the clean water change? Was the final pH of the polluted water the same as the pH of the clean water before cleaning? Explain.

Results Table

	Before cleaning (clean water)	Before cleaning (polluted water)	After filtration (clean water)	After filtration (polluted water)	After evaporation (clean water)	After evaporation (polluted water)
Color						
Particles						
Odor						
Suds						
pH						

DO NOT WRITE IN BOOK

Preparation Notes

Varying the thickness of the layers can contribute to a variation in the results. Specify the thickness of each layer so that the results between groups are comparable. The layers could be the following dimensions: 7 cm of gravel, 2.5 cm of charcoal, 10 cm of sand, and 10 cm of gravel. You may adjust the layers according to your class size or the size of the bottles.

For variation, try different sizes of gravel or different textures of sand. Both will affect how many particles travel through the filter. If you place a few drops of food coloring in the water, students can watch the progress of the water as it passes through the filter.

You may decide that it is easier and safer for you to perform step 2 of Part A, cutting the soda bottles, before class. You may also wish to perform steps 2–5 of Part B ahead of time. Some students may find it difficult to attach the plastic tubing to the glass tube or to slide the glass tube into the rubber stopper.

CHAPTER RESOURCES

Chapter Resource File

- Datasheet for LabBook
- Lab Notes and Answers

Part A
Analyze the Results

1. Sample answer: The filtered water was lighter in color than the unfiltered water. The water was still not as clear as the clean water; The color of the clean water stayed about the same.

2. Sample answer: no; The filtration method did not remove all of the particles from the polluted water; Many of the particles passed through the filter, but there were fewer particles than before the filtration.

3. Sample answer: The pH of the water changed slightly; The final pH of the polluted water was not the same as that of the clean water. After the polluted water was filtered, its pH was still slightly more acidic than that of the clean water.

Part B
Analyze the Results

1. Sample answer: After the evaporation process, the polluted water was clear. The color of the clean water did not change.

2. Sample answer: No particles were visible in the treated water after the evaporation method; The particles in the polluted water were left behind when the water evaporated.

3. Answers may vary, depending on the clean-water source. Generally, the pH of the two samples should be very close or the same.

Part B: Evaporation

Cleaning water by evaporation is more expensive than cleaning water by filtration. Evaporation requires more energy, which can come from a variety of sources. In this activity, you will use an electric hot plate as the energy source. See how well this method works to clean your sample of polluted water.

Form a Hypothesis

1 Write a hypothesis about which method you think will work better for water purification. Explain your reasoning.

Test the Hypothesis

2 Fill an Erlenmeyer flask with about 250 mL of the clean water, and insert the rubber stopper and glass tube into the flask.

3 Wearing goggles and gloves, connect about 1.5 m of plastic tubing to the glass tube.

4 Set the flask on the hot plate, and run the plastic tubing up and around the ring and down into a clean, empty 400 mL collection beaker.

5 Fill the sandwich bag with ice, seal the bag, and place the bag on the ring stand. Be sure the plastic bag and the tubing touch, as shown below.

6 Bring the water in the flask to a slow boil. As the water vapor passes by the bag of ice, the vapor will condense and drip into the collection beaker.

7 Observe the water in the collection beaker for color, particles, odor, suds, and pH. Record your observations in the "After evaporation" column in your data table.

8 Repeat steps 2–7 using the polluted water.

Analyze the Results

1 How did the color of the polluted water change after evaporation? Did the color of the clean water change after evaporation?

2 Did the evaporation method remove all of the particles from the polluted water? Explain.

3 How much did the pH of the polluted water change? Did the pH of the final clean water change? Was the final pH of the polluted water the same as the pH of the clean water before it was cleaned? Explain.

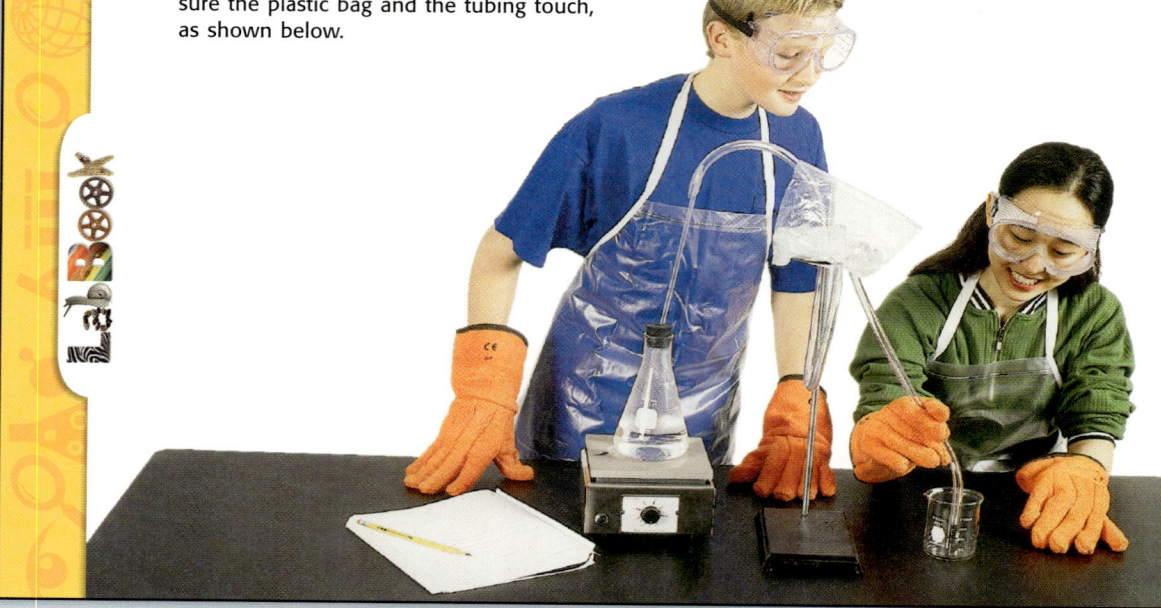

Draw Conclusions: Parts A and B

4 Which method—filtration or evaporation—removed the most pollutants from the water? Explain your reasoning.

5 Describe any changes that occurred in the clean water during this experiment.

6 What do you think are the advantages and disadvantages of each method?

7 Explain how you think each material (sand, gravel, and charcoal) used in the filtration system helped clean the water.

8 List areas of the country where you think each method of purification would be the most and the least beneficial. Explain your reasoning.

Applying Your Data

Do you think either purification method would remove oil from water? If time permits, repeat your experiment using several spoonfuls of cooking oil as the pollutant.

Filtration is only one step in the purification of water at water treatment plants. Research other methods used to purify public water supplies.

Parts A and B
Draw Conclusions

4. Sample answer: The evaporation method removed the most pollutants from the water; The pollutants were left behind when the water evaporated.

5. Sample answer: The clean water picked up some particles as it traveled through the filter. The water was not as clean after the filtration as it had been before the filtration.

6. Sample answer: The advantages of the filtration method are that: it is easy to do, it can treat large amounts of water, and it works relatively quickly. The filtration method doesn't remove all of the pollutants, however.
   The evaporation method removed more of the pollutants, but it is time consuming and expensive, particularly with large volumes of water.

7. Sample answer: The sand filtered out some of the larger particles and some of the soap. The gravel also removed the large particles. The charcoal removed most of the smaller particles, the odors, and some of the soap.

8. Answers may vary.

Applying Your Data

- Oil is removed from the water in both methods. The filtration method is relatively quicker than the evaporation method.

- Other methods of water purification include reverse osmosis, settling ponds, chemical additives such as chlorine, and other filtration layers. Students may also find out about bioremediation and the use of bacteria and plants to clean polluted water.

Investigating an Oil Spill

Teacher's Notes

Time Required

One 45-minute class period

Lab Ratings

EASY ————————————— HARD

Teacher Prep 🧪🧪🧪
Student Set-Up 🧪
Concept Level 🧪🧪
Clean Up 🧪🧪🧪

MATERIALS

The materials listed on the student page are sufficient for a group of 2 to 4 students. You may also choose to perform this activity as a demonstration to limit the use of machine oil.

Safety Caution

Remind students to review all safety cautions and icons before beginning this lab activity. Machine oil releases a strong odor. Use it only in well-ventilated areas.

Preparation Notes

Light machine oil may be replaced by cooking oil to demonstrate the same principle. If you choose to use machine oil, be sure there is sufficient ventilation in your classroom.

Investigating an Oil Spill

Have you ever wondered why it is important to recycle motor oil rather than pour it down the drain or sewer? Or have you ever wondered why a seemingly small oil spill can cause so much damage? The reason is that a little oil goes a long way.

Observing Oil and Water

Maybe you've heard the phrase "Oil and water don't mix." Oil dropped in water will spread out thinly over the surface of the water. In this activity, you'll learn how far a drop of oil can spread.

Ask a Question

1 How far will one drop of oil spread in a pan of water?

Form a Hypothesis

2 Write a hypothesis that could answer the question above.

Test the Hypothesis

3 Use a pipet to place one drop of oil into the middle of a pan of water. **Caution:** Machine oil is poisonous. Wear goggles and gloves. Keep materials that have contacted oil out of your mouth and eyes.

4 Observe what happens to the drop of oil for the next few seconds. Record your observations.

5 Using a metric ruler, measure the diameter of the oil slick to the nearest centimeter.

6 Determine the area of the oil slick in square centimeters. Use the formula below to find the area of a circle ($A = \pi r^2$). The radius (r) is equal to the diameter you measured in step 5 divided by 2. Multiply the radius by itself to get the square of the radius (r^2). Pi (π) is equal to 3.14. Record your answer.

> **Example**
>
> If your diameter is 10 cm,
>
> $r = 5$ cm, $r^2 = 25$ cm^2, $\pi = 3.14$
>
> $A = \pi r^2$
>
> $A = 3.14 \times 25$ cm^2
>
> $A = 78.5$ cm^2

MATERIALS

- calculator (optional)
- gloves, protective
- goggles
- graduated cylinder
- oil, light machine, 15 mL
- pan, large, at least 22 cm in diameter
- pipet
- ruler, metric
- water

SAFETY

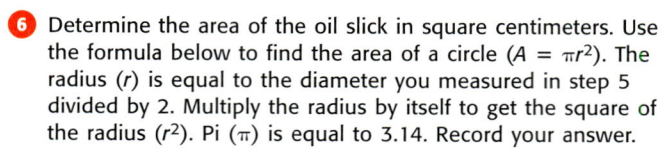

David Sparks
Redwater Jr. High
Redwater, Texas

Analyze the Results

1 What happened to the drop of oil when it came in contact with the water?

2 What total surface area was covered by the oil slick? (Show your calculations.)

Draw Conclusions

3 What can you conclude about the density of oil compared with the density of water?

Finding the Number of Drops in a Liter

"It's only a few drops," you may think as you spill something toxic on the ground. But those drops eventually add up. Just how many drops does it take to make a difference? In this activity, you'll learn just what an impact a few drops can have.

Procedure

1 Using a clean pipet, count the number of water drops it takes to fill the graduated cylinder to 10 mL. Be sure to add the drops slowly so you get an accurate count.

2 Since there are 1,000 mL in a liter, multiply the number of drops in 10 mL by 100. The result is the number of drops in a liter.

Analyze the Results

1 How many drops of water from your pipet did it take to fill a 1 L container?

2 What would happen if someone spilled 4 L of oil into a lake?

Applying Your Data

Can you devise a way to clean the oil from the water? Get permission from your teacher before testing your cleaning method.

Do you think oil behaves the same way in ocean water? Devise an experiment to test your hypothesis.

Analyze the Results

1. When the drop of oil touched the water, the oil spread out. This may surprise some students, who might expect the two substances to mix or the oil to sink.

2. Answers may vary. Students should show their work to illustrate that they understand the mathematical principles involved.

Draw Conclusions

3. Sample answer: This experiment shows that oil is less dense than water. For this reason, oil floats on water.

Analyze the Results

1. Answers may vary. The answer depends on the number of drops the students count in 10 mL of water. They should show their work to illustrate that they understand the mathematical principles involved.

2. The oil would spread to cover a large area. Students will not be able to determine the exact area for such an oil slick, but they should understand that the oil will spread significantly and pollute much of the lake.

Applying Your Data

• Answers may vary. One possibility is to use detergent to change the surface tension and remove the oil.

• To test the behavior of oil in ocean water, students could repeat the activity, using cold salt water instead of tap water. They could also rock the pan back and forth to simulate waves.

CHAPTER RESOURCES

Chapter Resource File

• Datasheet for LabBook
• Lab Notes and Answers

Disposal Information

Always follow federal, state, and local guidelines when disposing of oil. Pour cooking oil into a container of sand, and put it in the trash.

Model-Making Lab

Turning the Tides

Teacher's Notes

Time Required
One 45-minute class period

Lab Ratings

EASY ———————————— HARD

Teacher Prep 🧪🧪
Student Set-Up 🧪🧪🧪
Concept Level 🧪🧪🧪
Clean Up 🧪

MATERIALS

The materials listed on the student page are enough for a group of 2 to 4 students.

Safety Caution

Remind students to review all safety cautions and icons before beginning this lab activity. Students should wear safety goggles. Be sure that students have enough space to spin the system.

Model-Making Lab

Turning the Tides

Daily tides are caused by two "bulges" on the ocean's surface—one on the side of the Earth facing the moon and the other on the opposite side of the Earth. The bulge on the side facing the moon is caused by the moon's gravitational pull on the water. But the bulge on the opposite side of the Earth is slightly more difficult to explain. Whereas the moon pulls the water on one side of the Earth, the combined rotation of the Earth and the moon "pushes" the water on the opposite side of the Earth. In this activity, you will model the motion of the Earth and the moon to investigate the tidal bulge on the side of Earth facing away from the moon.

MATERIALS

- cardboard, 1 cm × 1 cm piece
- corrugated cardboard, one large and one small, with centers marked (2 disks)
- dowel, $\frac{1}{4}$ in. in diameter and 36 cm long
- glue, white
- pencil, sharp
- stapler with staples
- string, 5 cm length

SAFETY

Procedure

1 Draw a line from the center of each disk along the folds in the cardboard to the edge of the disk. This line is the radius.

2 Place a drop of white glue on one end of the dowel. Lay the larger disk flat, and align the dowel with the line for the radius you drew in step 1. Insert about 2.5 cm of the dowel into the edge of the disk.

3 Add a drop of glue to the other end of the dowel, and push that end into the smaller disk, again along its radius. The setup should look like a large, two-headed lollipop, as shown below. This setup is a model of the Earth-moon system.

4 Staple the string to the edge of the large disk on the side opposite the dowel. Staple the cardboard square to the other end of the string. This smaller piece of cardboard represents the Earth's oceans that face away from the moon.

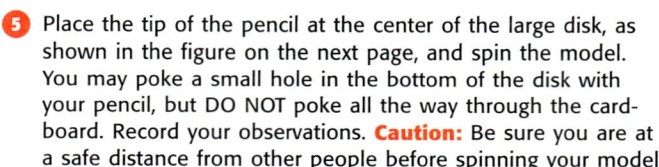

5 Place the tip of the pencil at the center of the large disk, as shown in the figure on the next page, and spin the model. You may poke a small hole in the bottom of the disk with your pencil, but DO NOT poke all the way through the cardboard. Record your observations. **Caution:** Be sure you are at a safe distance from other people before spinning your model.

Preparation Notes

You will need to put a mark at the center of all of the cardboard disks. Students will need to draw the radius of the circle from the center to the edge of the disk. Encourage students to draw this line along the corrugations of the cardboard. Otherwise, several other steps will be made more difficult.

The cardboard disks are not to scale with the Earth and moon. They are used to show how a two-body system, such as the Earth-moon system, rotates. The disks must be different sizes. The large disks could be 10 cm in diameter, and the smaller disks could be 5 cm in diameter.

6 Now, find your model's center of mass. The center of mass is the point at which the model can be balanced on the end of the pencil. (Hint: It might be easier to find the center of mass by using the eraser end. Then, use the sharpened end of the pencil to balance the model.) This balance point should be just inside the edge of the larger disk.

7 Place the pencil at the center of mass, and spin the model around the pencil. Again, you may wish to poke a small hole in the disk. Record your observations.

Analyze the Results

1 What happened when you tried to spin the model around the center of the large disk? This model, called the Earth-centered model, represents the incorrect view that the moon orbits the center of the Earth.

2 What happened when you tried to spin the model around its center of mass? This point, called the *barycenter,* is the point around which both the Earth and the moon rotate.

3 In each case, what happened to the string and cardboard square when the model was spun?

Draw Conclusions

4 Which model—the Earth-centered model or the barycentric model—explains why the Earth has a tidal bulge on the side opposite the moon? Explain.

Moon

Earth

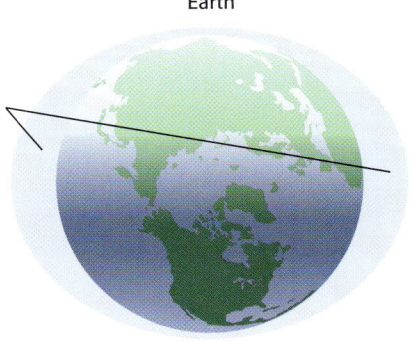
Tidal bulges

Analyze the Results

1. Answers may vary. Sample answer: When I tried to spin the model around the center of the large disk, I could not get the model to balance on the pencil.

2. Answers may vary. Sample answer: I was able to balance the model at the barycenter. The model spun, and the small piece of cardboard on the string swung outward.

3. Answers may vary. Sample answer: The cardboard square hung down when I tried to swing the model around the center of the large disk. I was unable to make the model spin. When I spun the model around its barycenter, the square swung away from the model.

Draw Conclusions

4. Sample answer: The barycentric model explains why the Earth has a bulge on the side opposite the moon. The side of the Earth opposite the side facing the moon acts in much the same way the small square of cardboard does in this model. As the Earth-moon system rotates, the side of the Earth facing away from the moon bulges outward.

CLASSROOM TESTED & APPROVED

Tracy Jahn
Berkshire Jr.-Sr. High
Canaan, New York

CHAPTER RESOURCES

Chapter Resource File
- Datasheet for LabBook
- Lab Notes and Answers

✓ Reading Check Answers

Chapter 1 The Flow of Fresh Water

Section 1

Page 4: The Colorado River eroded the rock over millions of years.

Page 6: A divide is the boundary that separates drainage areas, whereas a watershed is the area of land that is drained by a water system.

Page 7: An increase in a stream's gradient and discharge can cause the stream to flow faster.

Page 9: A mature river erodes its channel wider rather than deeper. It is not steep and has fewer falls and rapids. It also has good drainage and more discharge than a youthful river does.

Page 10: Rejuvenated rivers form when the land is raised by tectonic forces.

Section 2

Page 13: Deltas are made of the deposited load of the river, which is mostly mud.

Page 15: The flow of water can be controlled by dams and levees.

Section 3

Page 16: The zone of aeration is located underground. It is the area above the water table.

Page 18: The size of the recharge zone depends on how permeable rock is at the surface.

Page 19: A well must be deeper than the water table for it to be able to reach water.

Page 20: Deposition is the process that causes the formation of stalactites and stalagmites.

Section 4

Page 22: Nonpoint-source pollution is the hardest to control.

Page 25: Less than 8% of water in our homes is used for drinking.

Page 26: Drip irrigation systems deliver small amounts of water directly to the roots of the plant so that the plant absorbs the water before it can evaporate or runoff.

Page 27: Answers may vary. Sample answer: taking shorter showers, avoiding running water while brushing your teeth, and using the dishwasher only when it is full.

Chapter 2 Exploring the Oceans

Section 1

Page 39: The first oceans began to form sometime before 4 billion years ago as the Earth cooled enough for water vapor to condense and fall as rain.

Page 40: Coastal water in places with hotter, drier climates has a higher salinity because less fresh water runs into the ocean in drier areas and because heat increases the evaporation rate.

Page 42: Parts of the ocean along the equator are warmer because they receive more sunlight per year.

Page 44: If the ocean did not release thermal energy so slowly, the air temperature on land would vary greatly from above 100°C during the day to below 100°C at night.

Section 2

Page 47: Satellite photos from *Seasat* send images of the ocean back to Earth. These images allow scientists to measure the direction and speed of ocean currents. Satellite photos and information from *Geosat* have been used to measure slight changes in the height of the ocean's surface.

Page 48: 64,000 km; on the ocean floor

Page 49: continental shelf, continental slope, and continental rise

Page 50: It is unique because some organisms living around the vent do not rely on photosynthesis for energy.

Section 3

Page 53: The tough shells of clams and oysters protect the organisms against strong waves and harsh sunlight.

Page 55: crabs, sponges, worms, and sea cucumbers

Page 56: The neritic zone contains the largest concentration of marine life in the ocean because it receives more sunlight than the other zones in the ocean.

Section 4

Page 59: Fish farms can help reduce overfishing because the fish are raised instead of fished directly out of the ocean.

Page 60: Nonrenewable resources are resources that cannot be replenished. Oil and natural gas are nonrenewable resources.

Page 61: Desalination plants are most likely to be built in drier parts of the world, and where governments can afford to buy expensive equipment. Most desalination plants are in the Middle East, where the fuel needed to run the plants is relatively inexpensive.

Page 63: Wave energy would be a good alternative energy resource because it is a clean and renewable resource.

Section 5

Page 65: One effect of trash dumping is that plastic materials may harm and kill marine animals because these animals may mistake the trash for food.

Page 67: An oil tanker that has two hulls can prevent an oil spill, because if the outer hull is damaged, the inner hull will prevent oil from spilling into the ocean.

Page 69: The U.S. Marine Protection, Research, and Sanctuaries Act prohibits the dumping of any material that would affect human health or welfare, the marine environment or ecosystems, or businesses that depend on the ocean.

Chapter 3   The Movement of Ocean Water

Section 1

Page 80: Heyerdahl theorized that the inhabitants of Polynesia originally sailed from Peru on rafts powered only by the wind and ocean currents. Heyerdahl proved his theory by sailing from Peru to Polynesia on a raft powered only by wind and ocean currents.

Page 82: The Earth's rotation causes surface currents to move in curved paths rather than in straight lines.

Page 83: The three factors that form a pattern of surface currents on Earth are global winds, the Coriolis effect, and continental deflections.

Page 84: Density causes variations in the movement of deep currents.

Section 2

Page 87: Cold-water currents keep coastal climates cooler than inland climates all year long.

Page 89: Answers may vary. Sample answer: It is important to study El Niño because El Niño can greatly affect organisms and land. One way that scientists study El Niño is through a network of buoys located along the equator. These buoys record information that helps scientists predict when an El Niño is likely to occur.

Section 3

Page 90: The lowest point of a wave is called a *trough*.

Page 92: Deep-water waves become shallow-water waves as they move toward the shore and reach water that is shallower than one-half their wavelength.

Page 95: A storm surge is a local rise in sea level near the shore and is caused by strong winds from a storm, such as a hurricane. Storm surges are difficult to study because they disappear as quickly as they form.

Section 4

Page 96: The gravity of the moon pulls on every particle of the Earth.

Page 98: A tidal range is the difference between levels of ocean water at high tide and low tide.

Study Skills

FoldNote Instructions

Have you ever tried to study for a test or quiz but didn't know where to start? Or have you read a chapter and found that you can remember only a few ideas? Well, FoldNotes are a fun and exciting way to help you learn and remember the ideas you encounter as you learn science!

FoldNotes are tools that you can use to organize concepts. By focusing on a few main concepts, FoldNotes help you learn and remember how the concepts fit together. They can help you see the "big picture." Below you will find instructions for building 10 different FoldNotes.

Pyramid

1. Place a sheet of paper in front of you. Fold the lower left-hand corner of the paper diagonally to the opposite edge of the paper.

2. Cut off the tab of paper created by the fold (at the top).

3. Open the paper so that it is a square. Fold the lower right-hand corner of the paper diagonally to the opposite corner to form a triangle.

4. Open the paper. The creases of the two folds will have created an X.

5. Using scissors, cut along one of the creases. Start from any corner, and stop at the center point to create two flaps. Use tape or glue to attach one of the flaps on top of the other flap.

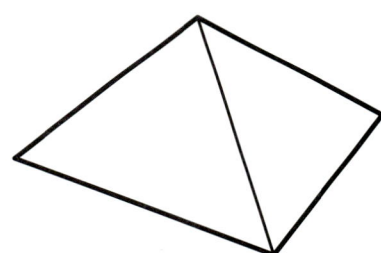

Double Door

1. Fold a sheet of paper in half from the top to the bottom. Then, unfold the paper.

2. Fold the top and bottom edges of the paper to the crease.

Booklet

1. Fold a sheet of paper in half from left to right. Then, unfold the paper.

2. Fold the sheet of paper in half again from the top to the bottom. Then, unfold the paper.

3. Refold the sheet of paper in half from left to right.

4. Fold the top and bottom edges to the center crease.

5. Completely unfold the paper.

6. Refold the paper from top to bottom.

7. Using scissors, cut a slit along the center crease of the sheet from the folded edge to the creases made in step 4. Do not cut the entire sheet in half.

8. Fold the sheet of paper in half from left to right. While holding the bottom and top edges of the paper, push the bottom and top edges together so that the center collapses at the center slit. Fold the four flaps to form a four-page book.

Layered Book

1. Lay one sheet of paper on top of another sheet. Slide the top sheet up so that 2 cm of the bottom sheet is showing.

2. Hold the two sheets together, fold down the top of the two sheets so that you see four 2 cm tabs along the bottom.

3. Using a stapler, staple the top of the FoldNote.

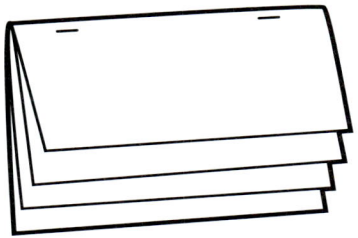

Key-Term Fold

1. Fold a sheet of lined notebook paper in half from left to right.

2. Using scissors, cut along every third line from the right edge of the paper to the center fold to make tabs.

Four-Corner Fold

1. Fold a sheet of paper in half from left to right. Then, unfold the paper.

2. Fold each side of the paper to the crease in the center of the paper.

3. Fold the paper in half from the top to the bottom. Then, unfold the paper.

4. Using scissors, cut the top flap creases made in step 3 to form four flaps.

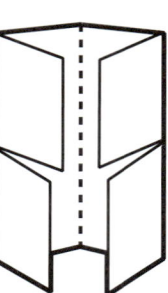

Three-Panel Flip Chart

1. Fold a piece of paper in half from the top to the bottom.

2. Fold the paper in thirds from side to side. Then, unfold the paper so that you can see the three sections.

3. From the top of the paper, cut along each of the vertical fold lines to the fold in the middle of the paper. You will now have three flaps.

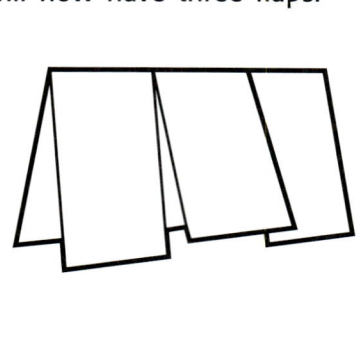

Table Fold

1. Fold a piece of paper in half from the top to the bottom. Then, fold the paper in half again.

2. Fold the paper in thirds from side to side.

3. Unfold the paper completely. Carefully trace the fold lines by using a pen or pencil.

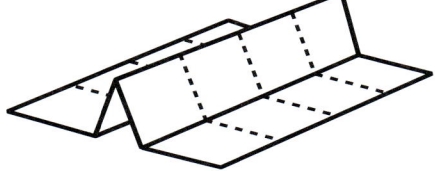

Two-Panel Flip Chart

1. Fold a piece of paper in half from the top to the bottom.

2. Fold the paper in half from side to side. Then, unfold the paper so that you can see the two sections.

3. From the top of the paper, cut along the vertical fold line to the fold in the middle of the paper. You will now have two flaps.

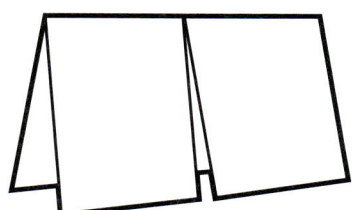

Tri-Fold

1. Fold a piece a paper in thirds from the top to the bottom.

2. Unfold the paper so that you can see the three sections. Then, turn the paper sideways so that the three sections form vertical columns.

3. Trace the fold lines by using a pen or pencil. Label the columns "Know," "Want," and "Learn."

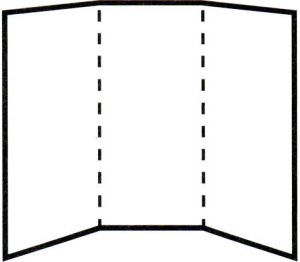

Graphic Organizer Instructions

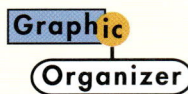 Have you ever wished that you could "draw out" the many concepts you learn in your science class? Sometimes, being able to *see* how concepts are related really helps you remember what you've learned. Graphic Organizers do just that! They give you a way to draw or map out concepts.

All you need to make a Graphic Organizer is a piece of paper and a pencil. Below you will find instructions for four different Graphic Organizers designed to help you organize the concepts you'll learn in this book.

Spider Map

1. Draw a diagram like the one shown. In the circle, write the main topic.

2. From the circle, draw legs to represent different categories of the main topic. You can have as many categories as you want.

3. From the category legs, draw horizontal lines. As you read the chapter, write details about each category on the horizontal lines.

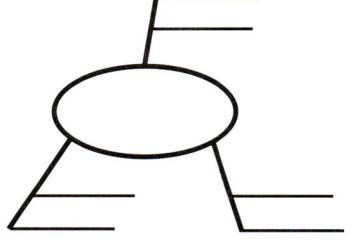

Comparison Table

1. Draw a chart like the one shown. Your chart can have as many columns and rows as you want.

2. In the top row, write the topics that you want to compare.

3. In the left column, write characteristics of the topics that you want to compare. As you read the chapter, fill in the characteristics for each topic in the appropriate boxes.

Chain-of-Events-Chart

1. Draw a box. In the box, write the first step of a process or the first event of a timeline.

2. Under the box, draw another box, and use an arrow to connect the two boxes. In the second box, write the next step of the process or the next event in the timeline.

3. Continue adding boxes until the process or timeline is finished.

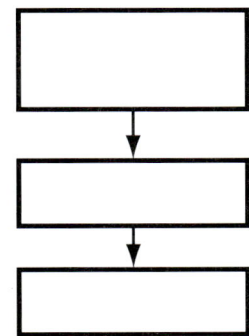

Concept Map

1. Draw a circle in the center of a piece of paper. Write the main idea of the chapter in the center of the circle.

2. From the circle, draw other circles. In those circles, write characteristics of the main idea. Draw arrows from the center circle to the circles that contain the characteristics.

3. From each circle that contains a characteristic, draw other circles. In those circles, write specific details about the characteristic. Draw arrows from each circle that contains a characteristic to the circles that contain specific details. You may draw as many circles as you want.

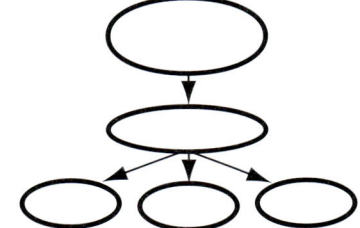

Appendix

SI Measurement

The International System of Units, or SI, is the standard system of measurement used by many scientists. Using the same standards of measurement makes it easier for scientists to communicate with one another.

SI works by combining prefixes and base units. Each base unit can be used with different prefixes to define smaller and larger quantities. The table below lists common SI prefixes.

SI Prefixes			
Prefix	Symbol	Factor	Example
kilo-	k	1,000	kilogram, 1 kg = 1,000 g
hecto-	h	100	hectoliter, 1 hL = 100 L
deka-	da	10	dekameter, 1 dam = 10 m
		1	meter, liter, gram
deci-	d	0.1	decigram, 1 dg = 0.1 g
centi-	c	0.01	centimeter, 1 cm = 0.01 m
milli-	m	0.001	milliliter, 1 mL = 0.001 L
micro-	μ	0.000 001	micrometer, 1 μm = 0.000 001 m

SI Conversion Table		
SI units	From SI to English	From English to SI
Length		
kilometer (km) = 1,000 m	1 km = 0.621 mi	1 mi = 1.609 km
meter (m) = 100 cm	1 m = 3.281 ft	1 ft = 0.305 m
centimeter (cm) = 0.01 m	1 cm = 0.394 in.	1 in. = 2.540 cm
millimeter (mm) = 0.001 m	1 mm = 0.039 in.	
micrometer (μm) = 0.000 001 m		
nanometer (nm) = 0.000 000 001 m		
Area		
square kilometer (km^2) = 100 hectares	1 km^2 = 0.386 mi^2	1 mi^2 = 2.590 km^2
hectare (ha) = 10,000 m^2	1 ha = 2.471 acres	1 acre = 0.405 ha
square meter (m^2) = 10,000 cm^2	1 m^2 = 10.764 ft^2	1 ft^2 = 0.093 m^2
square centimeter (cm^2) = 100 mm^2	1 cm^2 = 0.155 in.^{2}	1 in.^{2} = 6.452 cm^2
Volume		
liter (L) = 1,000 mL = 1 dm^3	1 L = 1.057 fl qt	1 fl qt = 0.946 L
milliliter (mL) = 0.001 L = 1 cm^3	1 mL = 0.034 fl oz	1 fl oz = 29.574 mL
microliter (μL) = 0.000 001 L		
Mass		
kilogram (kg) = 1,000 g	1 kg = 2.205 lb	1 lb = 0.454 kg
gram (g) = 1,000 mg	1 g = 0.035 oz	1 oz = 28.350 g
milligram (mg) = 0.001 g		
microgram (μg) = 0.000 001 g		

Appendix

Temperature Scales

Temperature can be expressed by using three different scales: Fahrenheit, Celsius, and Kelvin. The SI unit for temperature is the kelvin (K).

Although 0 K is much colder than 0°C, a change of 1 K is equal to a change of 1°C.

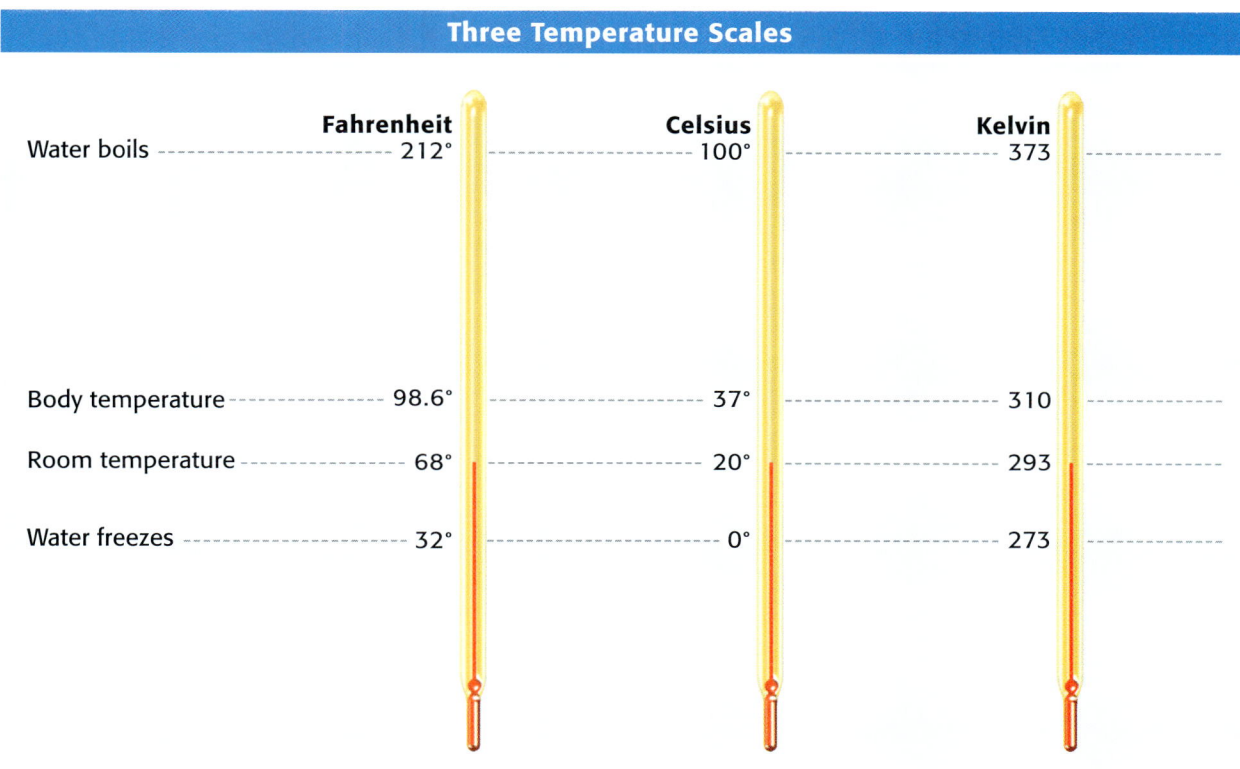

Three Temperature Scales

	Fahrenheit	Celsius	Kelvin
Water boils	212°	100°	373
Body temperature	98.6°	37°	310
Room temperature	68°	20°	293
Water freezes	32°	0°	273

Temperature Conversions Table

To convert	Use this equation:	Example
Celsius to Fahrenheit °C → °F	$°F = \left(\dfrac{9}{5} \times °C\right) + 32$	Convert 45°C to °F. $°F = \left(\dfrac{9}{5} \times 45°C\right) + 32 = 113°F$
Fahrenheit to Celsius °F → °C	$°C = \dfrac{5}{9} \times (°F - 32)$	Convert 68°F to °C. $°C = \dfrac{5}{9} \times (68°F - 32) = 20°C$
Celsius to Kelvin °C → K	$K = °C + 273$	Convert 45°C to K. $K = 45°C + 273 = 318\ K$
Kelvin to Celsius K → °C	$°C = K - 273$	Convert 32 K to °C. $°C = 32K - 273 = -241°C$

Measuring Skills

Using a Graduated Cylinder

When using a graduated cylinder to measure volume, keep the following procedures in mind:

1 Place the cylinder on a flat, level surface before measuring liquid.

2 Move your head so that your eye is level with the surface of the liquid.

3 Read the mark closest to the liquid level. On glass graduated cylinders, read the mark closest to the center of the curve in the liquid's surface.

Using a Meterstick or Metric Ruler

When using a meterstick or metric ruler to measure length, keep the following procedures in mind:

1 Place the ruler firmly against the object that you are measuring.

2 Align one edge of the object exactly with the 0 end of the ruler.

3 Look at the other edge of the object to see which of the marks on the ruler is closest to that edge. (Note: Each small slash between the centimeters represents a millimeter, which is one-tenth of a centimeter.)

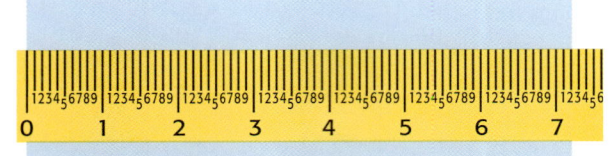

Using a Triple-Beam Balance

When using a triple-beam balance to measure mass, keep the following procedures in mind:

1 Make sure the balance is on a level surface.

2 Place all of the countermasses at 0. Adjust the balancing knob until the pointer rests at 0.

3 Place the object you wish to measure on the pan. **Caution:** Do not place hot objects or chemicals directly on the balance pan.

4 Move the largest countermass along the beam to the right until it is at the last notch that does not tip the balance. Follow the same procedure with the next-largest countermass. Then, move the smallest countermass until the pointer rests at 0.

5 Add the readings from the three beams together to determine the mass of the object.

6 When determining the mass of crystals or powders, first find the mass of a piece of filter paper. Then, add the crystals or powder to the paper, and remeasure. The actual mass of the crystals or powder is the total mass minus the mass of the paper. When finding the mass of liquids, first find the mass of the empty container. Then, find the combined mass of the liquid and container. The mass of the liquid is the total mass minus the mass of the container.

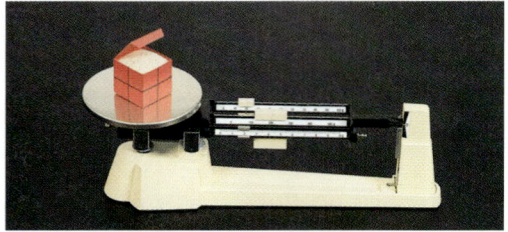

Scientific Methods

The ways in which scientists answer questions and solve problems are called **scientific methods.** The same steps are often used by scientists as they look for answers. However, there is more than one way to use these steps. Scientists may use all of the steps or just some of the steps during an investigation. They may even repeat some of the steps. The goal of using scientific methods is to come up with reliable answers and solutions.

Six Steps of Scientific Methods

 Good questions come from careful **observations.** You make observations by using your senses to gather information. Sometimes, you may use instruments, such as microscopes and telescopes, to extend the range of your senses. As you observe the natural world, you will discover that you have many more questions than answers. These questions drive investigations.

Questions beginning with *what, why, how,* and *when* are important in focusing an investigation. Here is an example of a question that could lead to an investigation.

Question: How does acid rain affect plant growth?

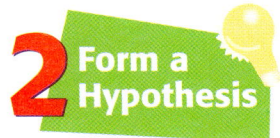 After you ask a question, you need to form a **hypothesis.** A hypothesis is a clear statement of what you expect the answer to your question to be. Your hypothesis will represent your best "educated guess" based on what you have observed and what you already know. A good hypothesis is testable. Otherwise, the investigation can go no further. Here is a hypothesis based on the question, "How does acid rain affect plant growth?"

Hypothesis: Acid rain slows plant growth.

The hypothesis can lead to predictions. A prediction is what you think the outcome of your experiment or data collection will be. Predictions are usually stated in an if-then format. Here is a sample prediction for the hypothesis that acid rain slows plant growth.

Prediction: If a plant is watered with only acid rain (which has a pH of 4), then the plant will grow at half its normal rate.

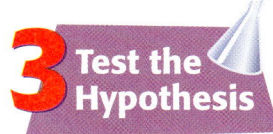 After you have formed a hypothesis and made a prediction, your hypothesis should be tested. One way to test a hypothesis is with a controlled experiment. A **controlled experiment** tests only one factor at a time. In an experiment to test the effect of acid rain on plant growth, the **control group** would be watered with normal rain water. The **experimental group** would be watered with acid rain. All of the plants should receive the same amount of sunlight and water each day. The air temperature should be the same for all groups. However, the acidity of the water will be a variable. In fact, any factor that is different from one group to another is a **variable.** If your hypothesis is correct, then the acidity of the water and plant growth are *dependant variables.* The amount a plant grows is dependent on the acidity of the water. However, the amount of water each plant receives and the amount of sunlight each plant receives are *independent variables.* Either of these factors could change without affecting the other factor.

Sometimes, the nature of an investigation makes a controlled experiment impossible. For example, the Earth's core is surrounded by thousands of meters of rock. Under such circumstances, a hypothesis may be tested by making detailed observations.

 After you have completed your experiments, made your observations, and collected your data, you must analyze all the information you have gathered. Tables and graphs are often used in this step to organize the data.

5 Draw Conclusions

After analyzing your data, you can determine if your results support your hypothesis. If your hypothesis is supported, you (or others) might want to repeat the observations or experiments to verify your results. If your hypothesis is not supported by the data, you may have to check your procedure for errors. You may even have to reject your hypothesis and make a new one. If you cannot draw a conclusion from your results, you may have to try the investigation again or carry out further observations or experiments.

6 Communicate Results

After any scientific investigation, you should report your results. By preparing a written or oral report, you let others know what you have learned. They may repeat your investigation to see if they get the same results. Your report may even lead to another question and then to another investigation.

Scientific Methods in Action

Scientific methods contain loops in which several steps may be repeated over and over again. In some cases, certain steps are unnecessary. Thus, there is not a "straight line" of steps. For example, sometimes scientists find that testing one hypothesis raises new questions and new hypotheses to be tested. And sometimes, testing the hypothesis leads directly to a conclusion. Furthermore, the steps in scientific methods are not always used in the same order. Follow the steps in the diagram, and see how many different directions scientific methods can take you.

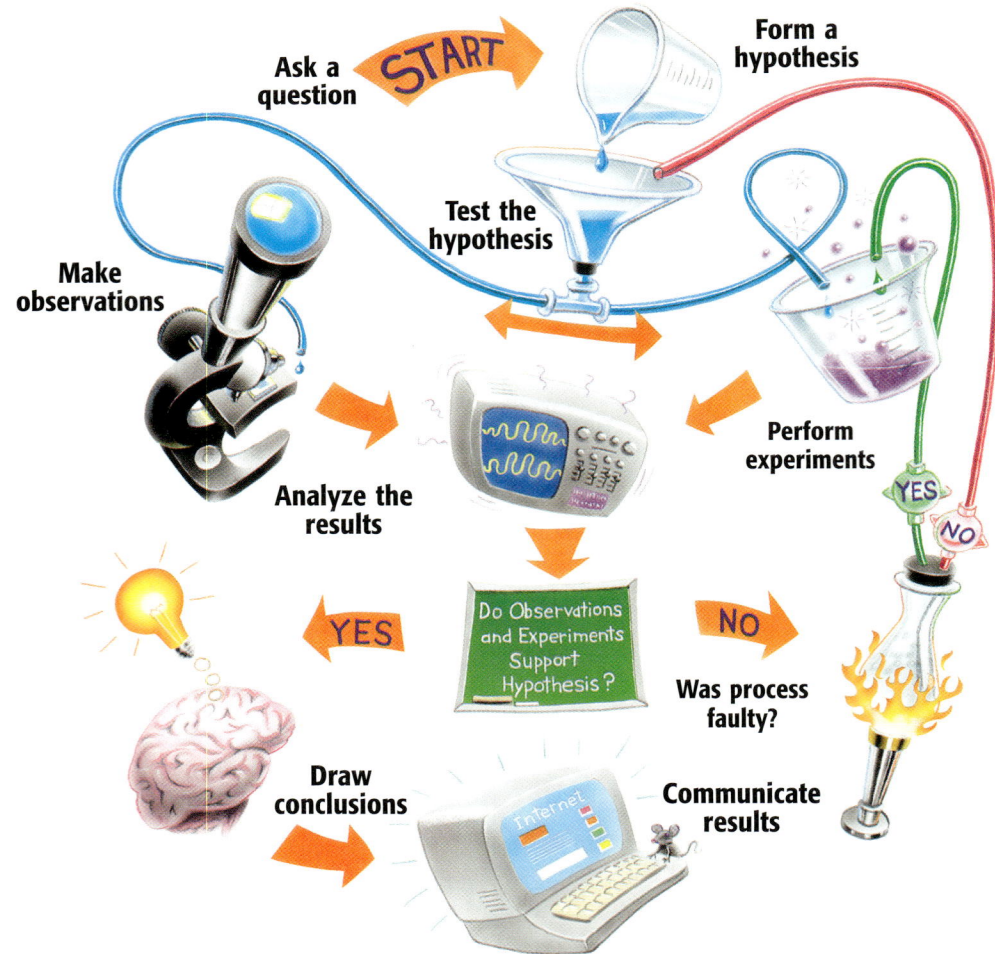

Appendix

Making Charts and Graphs

Pie Charts

A pie chart shows how each group of data relates to all of the data. Each part of the circle forming the chart represents a category of the data. The entire circle represents all of the data. For example, a biologist studying a hardwood forest in Wisconsin found that there were five different types of trees. The data table at right summarizes the biologist's findings.

Wisconsin Hardwood Trees	
Type of tree	**Number found**
Oak	600
Maple	750
Beech	300
Birch	1,200
Hickory	150
Total	3,000

How to Make a Pie Chart

1 To make a pie chart of these data, first find the percentage of each type of tree. Divide the number of trees of each type by the total number of trees, and multiply by 100.

$$\frac{600 \text{ oak}}{3,000 \text{ trees}} \times 100 = 20\%$$

$$\frac{750 \text{ maple}}{3,000 \text{ trees}} \times 100 = 25\%$$

$$\frac{300 \text{ beech}}{3,000 \text{ trees}} \times 100 = 10\%$$

$$\frac{1,200 \text{ birch}}{3,000 \text{ trees}} \times 100 = 40\%$$

$$\frac{150 \text{ hickory}}{3,000 \text{ trees}} \times 100 = 5\%$$

2 Now, determine the size of the wedges that make up the pie chart. Multiply each percentage by 360°. Remember that a circle contains 360°.

$20\% \times 360° = 72°$ $25\% \times 360° = 90°$

$10\% \times 360° = 36°$ $40\% \times 360° = 144°$

$5\% \times 360° = 18°$

3 Check that the sum of the percentages is 100 and the sum of the degrees is 360.

$20\% + 25\% + 10\% + 40\% + 5\% = 100\%$

$72° + 90° + 36° + 144° + 18° = 360°$

4 Use a compass to draw a circle and mark the center of the circle.

5 Then, use a protractor to draw angles of 72°, 90°, 36°, 144°, and 18° in the circle.

6 Finally, label each part of the chart, and choose an appropriate title.

A Community of Wisconsin Hardwood Trees

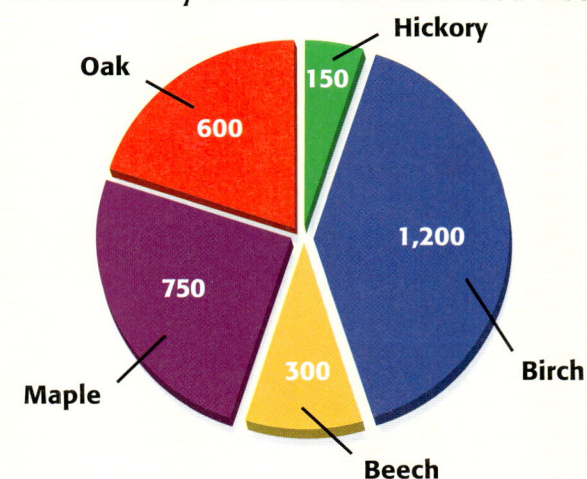

Line Graphs

Line graphs are most often used to demonstrate continuous change. For example, Mr. Smith's students analyzed the population records for their hometown, Appleton, between 1900 and 2000. Examine the data at right.

Because the year and the population change, they are the *variables*. The population is determined by, or dependent on, the year. Therefore, the population is called the **dependent variable**, and the year is called the **independent variable**. Each set of data is called a **data pair.** To prepare a line graph, you must first organize data pairs into a table like the one at right.

Population of Appleton, 1900–2000	
Year	Population
1900	1,800
1920	2,500
1940	3,200
1960	3,900
1980	4,600
2000	5,300

How to Make a Line Graph

1. Place the independent variable along the horizontal (*x*) axis. Place the dependent variable along the vertical (*y*) axis.

2. Label the *x*-axis "Year" and the *y*-axis "Population." Look at your largest and smallest values for the population. For the *y*-axis, determine a scale that will provide enough space to show these values. You must use the same scale for the entire length of the axis. Next, find an appropriate scale for the *x*-axis.

3. Choose reasonable starting points for each axis.

4. Plot the data pairs as accurately as possible.

5. Choose a title that accurately represents the data.

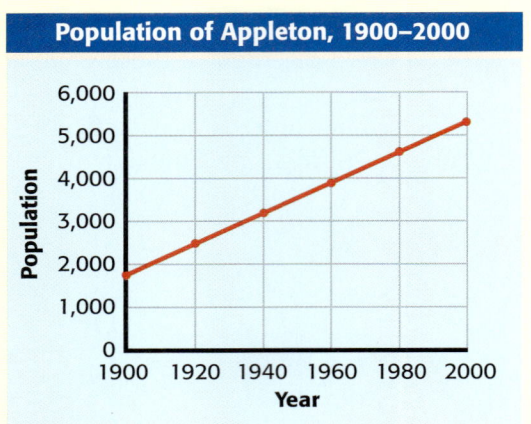

How to Determine Slope

Slope is the ratio of the change in the *y*-value to the change in the *x*-value, or "rise over run."

1. Choose two points on the line graph. For example, the population of Appleton in 2000 was 5,300 people. Therefore, you can define point *a* as (2000, 5,300). In 1900, the population was 1,800 people. You can define point *b* as (1900, 1,800).

2. Find the change in the *y*-value.
(*y* at point *a*) − (*y* at point *b*) = 5,300 people − 1,800 people = 3,500 people

3. Find the change in the *x*-value.
(*x* at point *a*) − (*x* at point *b*) = 2000 − 1900 = 100 years

4. Calculate the slope of the graph by dividing the change in *y* by the change in *x*.

$$slope = \frac{change\ in\ y}{change\ in\ x}$$

$$slope = \frac{3,500\ people}{100\ years}$$

$$slope = 35\ people\ per\ year$$

In this example, the population in Appleton increased by a fixed amount each year. The graph of these data is a straight line. Therefore, the relationship is **linear.** When the graph of a set of data is not a straight line, the relationship is **nonlinear.**

Using Algebra to Determine Slope

The equation in step 4 may also be arranged to be

$$y = kx$$

where y represents the change in the y-value, k represents the slope, and x represents the change in the x-value.

$$slope = \frac{change\ in\ y}{change\ in\ x}$$

$$k = \frac{y}{x}$$

$$k \times x = \frac{y \times x}{x}$$

$$kx = y$$

Bar Graphs

Bar graphs are used to demonstrate change that is not continuous. These graphs can be used to indicate trends when the data cover a long period of time. A meteorologist gathered the precipitation data shown here for Hartford, Connecticut, for April 1–15, 1996, and used a bar graph to represent the data.

Precipitation in Hartford, Connecticut April 1–15, 1996			
Date	Precipitation (cm)	Date	Precipitation (cm)
April 1	0.5	April 9	0.25
April 2	1.25	April 10	0.0
April 3	0.0	April 11	1.0
April 4	0.0	April 12	0.0
April 5	0.0	April 13	0.25
April 6	0.0	April 14	0.0
April 7	0.0	April 15	6.50
April 8	1.75		

How to Make a Bar Graph

1 Use an appropriate scale and a reasonable starting point for each axis.

2 Label the axes, and plot the data.

3 Choose a title that accurately represents the data.

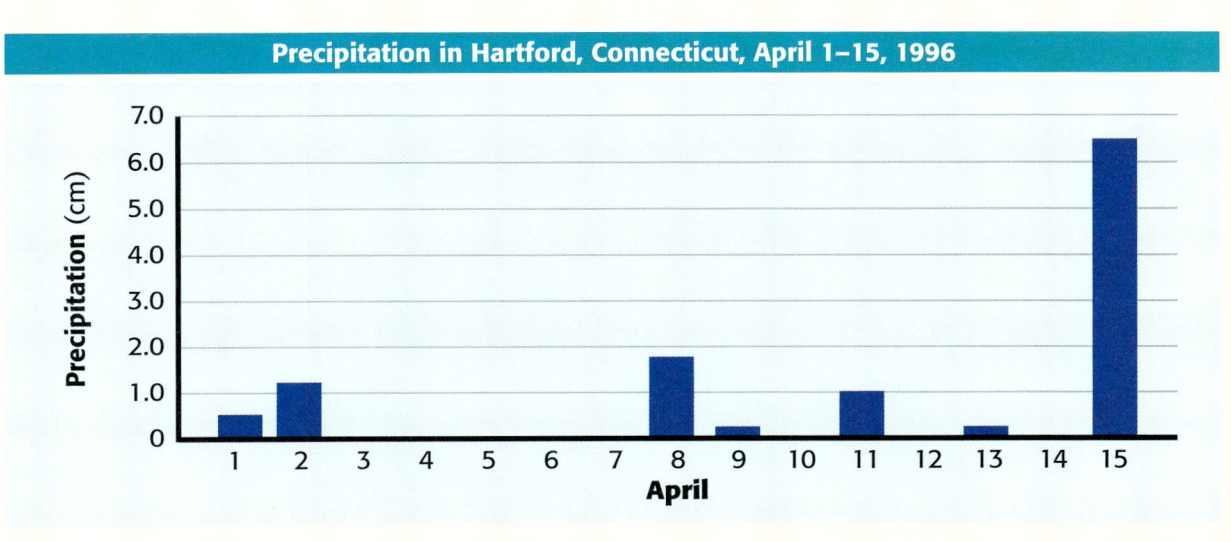

Precipitation in Hartford, Connecticut, April 1–15, 1996

Math Refresher

Science requires an understanding of many math concepts. The following pages will help you review some important math skills.

Averages

An **average**, or **mean,** simplifies a set of numbers into a single number that *approximates* the value of the set.

> **Example:** Find the average of the following set of numbers: 5, 4, 7, and 8.

Step 1: Find the sum.
$$5 + 4 + 7 + 8 = 24$$

Step 2: Divide the sum by the number of numbers in your set. Because there are four numbers in this example, divide the sum by 4.
$$\frac{24}{4} = 6$$

The average, or mean, is **6.**

Ratios

A **ratio** is a comparison between numbers, and it is usually written as a fraction.

> **Example:** Find the ratio of thermometers to students if you have 36 thermometers and 48 students in your class.

Step 1: Make the ratio.
$$\frac{36 \text{ thermometers}}{48 \text{ students}}$$

Step 2: Reduce the fraction to its simplest form.
$$\frac{36}{48} = \frac{36 \div 12}{48 \div 12} = \frac{3}{4}$$

The ratio of thermometers to students is **3 to 4,** or $\frac{3}{4}$. The ratio may also be written in the form 3:4.

Proportions

A **proportion** is an equation that states that two ratios are equal.
$$\frac{3}{1} = \frac{12}{4}$$

To solve a proportion, first multiply across the equal sign. This is called *cross-multiplication.* If you know three of the quantities in a proportion, you can use cross-multiplication to find the fourth.

> **Example:** Imagine that you are making a scale model of the solar system for your science project. The diameter of Jupiter is 11.2 times the diameter of the Earth. If you are using a plastic-foam ball that has a diameter of 2 cm to represent the Earth, what must the diameter of the ball representing Jupiter be?
> $$\frac{11.2}{1} = \frac{x}{2 \text{ cm}}$$

Step 1: Cross-multiply.
$$\frac{11.2}{1} \diagdown \frac{x}{2}$$
$$11.2 \times 2 = x \times 1$$

Step 2: Multiply.
$$22.4 = x \times 1$$

Step 3: Isolate the variable by dividing both sides by 1.
$$x = \frac{22.4}{1}$$
$$x = 22.4 \text{ cm}$$

You will need to use a ball that has a diameter of **22.4** cm to represent Jupiter.

Percentages

A **percentage** is a ratio of a given number to 100.

> **Example:** What is 85% of 40?

Step 1: Rewrite the percentage by moving the decimal point two places to the left.

$$0.85$$

Step 2: Multiply the decimal by the number that you are calculating the percentage of.

$$0.85 \times 40 = 34$$

85% of 40 is **34.**

Decimals

To **add** or **subtract decimals,** line up the digits vertically so that the decimal points line up. Then, add or subtract the columns from right to left. Carry or borrow numbers as necessary.

> **Example:** Add the following numbers: 3.1415 and 2.96.

Step 1: Line up the digits vertically so that the decimal points line up.

$$\begin{array}{r} 3.1415 \\ + 2.96 \\ \hline \end{array}$$

Step 2: Add the columns from right to left, and carry when necessary.

$$\begin{array}{r} {\scriptstyle 1\ \ 1} \\ 3.1415 \\ + 2.96 \\ \hline 6.1015 \end{array}$$

The sum is **6.1015.**

Fractions

Numbers tell you how many; **fractions** tell you *how much of a whole*.

> **Example:** Your class has 24 plants. Your teacher instructs you to put 5 plants in a shady spot. What fraction of the plants in your class will you put in a shady spot?

Step 1: In the denominator, write the total number of parts in the whole.

$$\frac{?}{24}$$

Step 2: In the numerator, write the number of parts of the whole that are being considered.

$$\frac{5}{24}$$

So, $\frac{5}{24}$ of the plants will be in the shade.

Reducing Fractions

It is usually best to express a fraction in its simplest form. Expressing a fraction in its simplest form is called *reducing* a fraction.

> **Example:** Reduce the fraction $\frac{30}{45}$ to its simplest form.

Step 1: Find the largest whole number that will divide evenly into both the numerator and denominator. This number is called the *greatest common factor* (GCF).

Factors of the numerator 30:
> 1, 2, 3, 5, 6, 10, **15,** 30

Factors of the denominator 45:
> 1, 3, 5, 9, **15,** 45

Step 2: Divide both the numerator and the denominator by the GCF, which in this case is 15.

$$\frac{30}{45} = \frac{30 \div 15}{45 \div 15} = \frac{2}{3}$$

Thus, $\frac{30}{45}$ reduced to its simplest form is $\frac{2}{3}$.

Adding and Subtracting Fractions

To **add** or **subtract fractions** that have the **same denominator,** simply add or subtract the numerators.

> **Examples:**
>
> $\frac{3}{5} + \frac{1}{5} = ?$ and $\frac{3}{4} - \frac{1}{4} = ?$

Step 1: Add or subtract the numerators.

$$\frac{3}{5} + \frac{1}{5} = \frac{4}{} \quad \text{and} \quad \frac{3}{4} - \frac{1}{4} = \frac{2}{}$$

Step 2: Write the sum or difference over the denominator.

$$\frac{3}{5} + \frac{1}{5} = \frac{4}{5} \quad \text{and} \quad \frac{3}{4} - \frac{1}{4} = \frac{2}{4}$$

Step 3: If necessary, reduce the fraction to its simplest form.

$\frac{4}{5}$ cannot be reduced, and $\frac{2}{4} = \frac{1}{2}$.

To **add** or **subtract fractions** that have **different denominators,** first find the least common denominator (LCD).

> **Examples:**
>
> $\frac{1}{2} + \frac{1}{6} = ?$ and $\frac{3}{4} - \frac{2}{3} = ?$

Step 1: Write the equivalent fractions that have a common denominator.

$$\frac{3}{6} + \frac{1}{6} = ? \quad \text{and} \quad \frac{9}{12} - \frac{8}{12} = ?$$

Step 2: Add or subtract the fractions.

$$\frac{3}{6} + \frac{1}{6} = \frac{4}{6} \quad \text{and} \quad \frac{9}{12} - \frac{8}{12} = \frac{1}{12}$$

Step 3: If necessary, reduce the fraction to its simplest form.

The fraction $\frac{4}{6} = \frac{2}{3}$, and $\frac{1}{12}$ cannot be reduced.

Multiplying Fractions

To **multiply fractions,** multiply the numerators and the denominators together, and then reduce the fraction to its simplest form.

> **Example:**
>
> $\frac{5}{9} \times \frac{7}{10} = ?$

Step 1: Multiply the numerators and denominators.

$$\frac{5}{9} \times \frac{7}{10} = \frac{5 \times 7}{9 \times 10} = \frac{35}{90}$$

Step 2: Reduce the fraction.

$$\frac{35}{90} = \frac{35 \div 5}{90 \div 5} = \frac{7}{18}$$

Dividing Fractions

To **divide fractions,** first rewrite the divisor (the number you divide by) upside down. This number is called the *reciprocal* of the divisor. Then multiply and reduce if necessary.

> **Example:**
>
> $\frac{5}{8} \div \frac{3}{2} = ?$

Step 1: Rewrite the divisor as its reciprocal.

$$\frac{3}{2} \rightarrow \frac{2}{3}$$

Step 2: Multiply the fractions.

$$\frac{5}{8} \times \frac{2}{3} = \frac{5 \times 2}{8 \times 3} = \frac{10}{24}$$

Step 3: Reduce the fraction.

$$\frac{10}{24} = \frac{10 \div 2}{24 \div 2} = \frac{5}{12}$$

Scientific Notation

Scientific notation is a short way of representing very large and very small numbers without writing all of the place-holding zeros.

Example: Write 653,000,000 in scientific notation.

Step 1: Write the number without the place-holding zeros.

653

Step 2: Place the decimal point after the first digit.

6.53

Step 3: Find the exponent by counting the number of places that you moved the decimal point.

6.53000000

The decimal point was moved eight places to the left. Therefore, the exponent of 10 is positive 8. If you had moved the decimal point to the right, the exponent would be negative.

Step 4: Write the number in scientific notation.

6.53×10^8

Area

Area is the number of square units needed to cover the surface of an object.

Formulas:

area of a square = side × side
area of a rectangle = length × width
area of a triangle = $\frac{1}{2}$ × base × height

Examples: Find the areas.

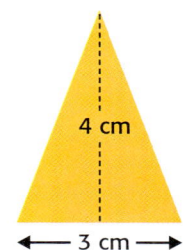

Triangle
area = $\frac{1}{2}$ × base × height
area = $\frac{1}{2}$ × 3 cm × 4 cm
*area = **6 cm²***

4 cm — 3 cm

Rectangle
area = length × width
area = 6 cm × 3 cm
*area = **18 cm²***

3 cm — 6 cm

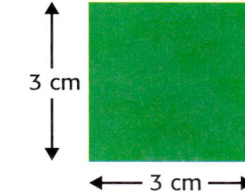

Square
area = side × side
area = 3 cm × 3 cm
*area = **9 cm²***

3 cm — 3 cm

Volume

Volume is the amount of space that something occupies.

Formulas:

*volume of a cube =
side × side × side*

*volume of a prism =
area of base × height*

Examples:

Find the volume of the solids.

Cube
volume = side × side × side
volume = 4 cm × 4 cm × 4 cm
*volume = **64 cm³***

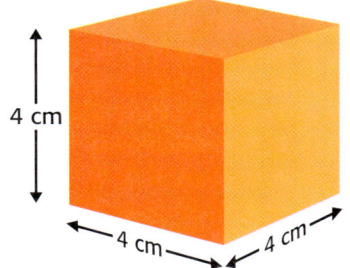

4 cm — 4 cm — 4 cm

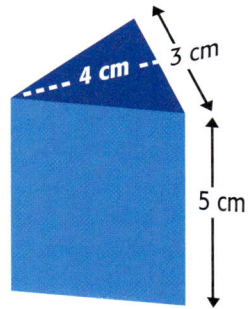

4 cm — 3 cm — 5 cm

Prism
volume = area of base × height
volume = (area of triangle) × height
volume = ($\frac{1}{2}$ × 3 cm × 4 cm) × 5 cm
volume = 6 cm² × 5 cm
*volume = **30 cm³***

Periodic Table of the Elements

Each square on the table includes an element's name, chemical symbol, atomic number, and atomic mass.

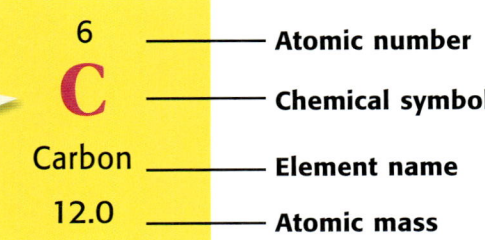

The color of the chemical symbol indicates the physical state at room temperature. Carbon is a solid.

6	——— Atomic number
C	——— Chemical symbol
Carbon	——— Element name
12.0	——— Atomic mass

The background color indicates the type of element. Carbon is a nonmetal.

Background
- Metals
- Metalloids
- Nonmetals

Chemical symbol
- Solid
- Liquid
- Gas

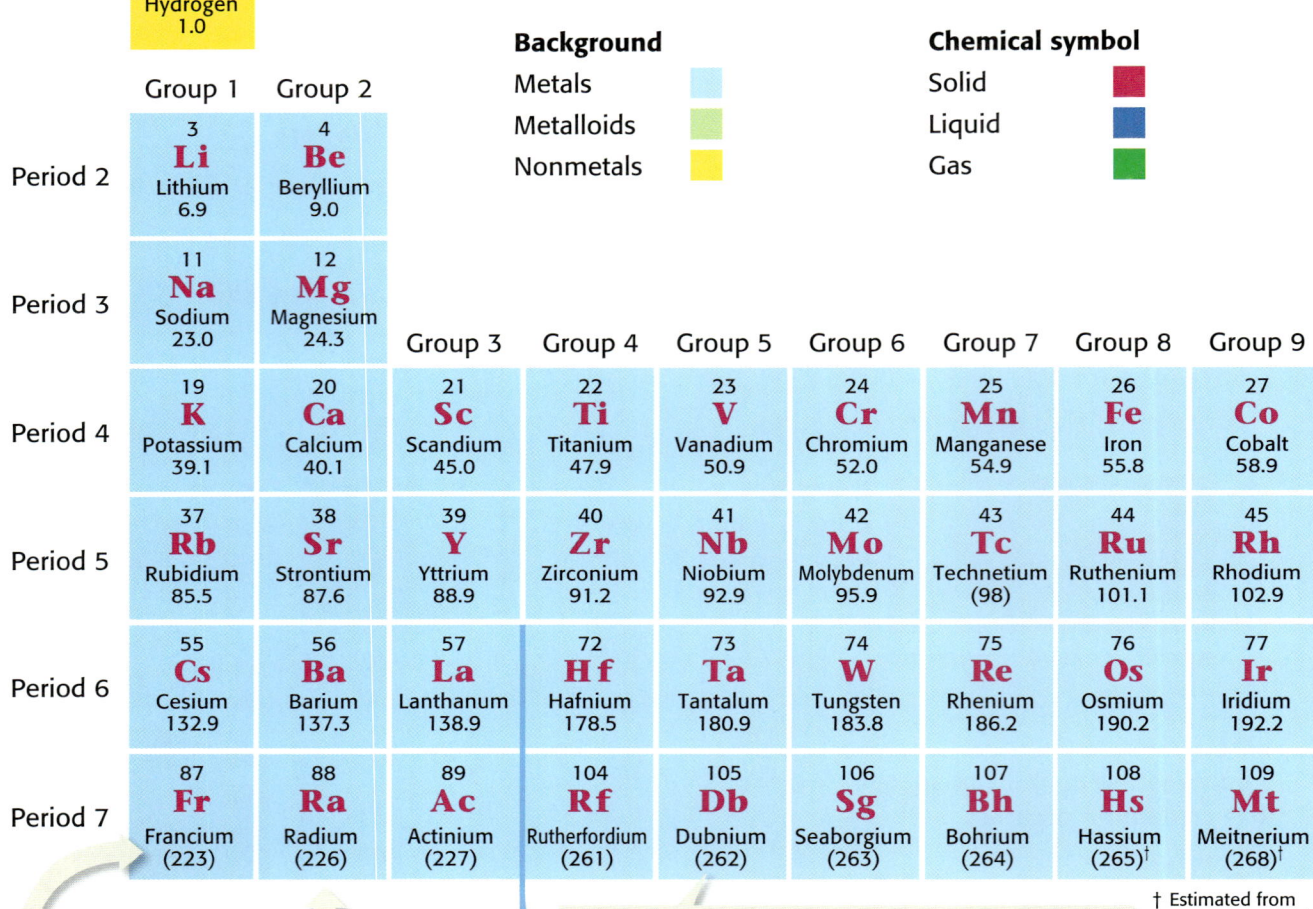

A row of elements is called a *period*.

A column of elements is called a *group* or *family*.

Values in parentheses are of the most stable isotope of the element.

† Estimated from currently available IUPAC data.

These elements are placed below the table to allow the table to be narrower.

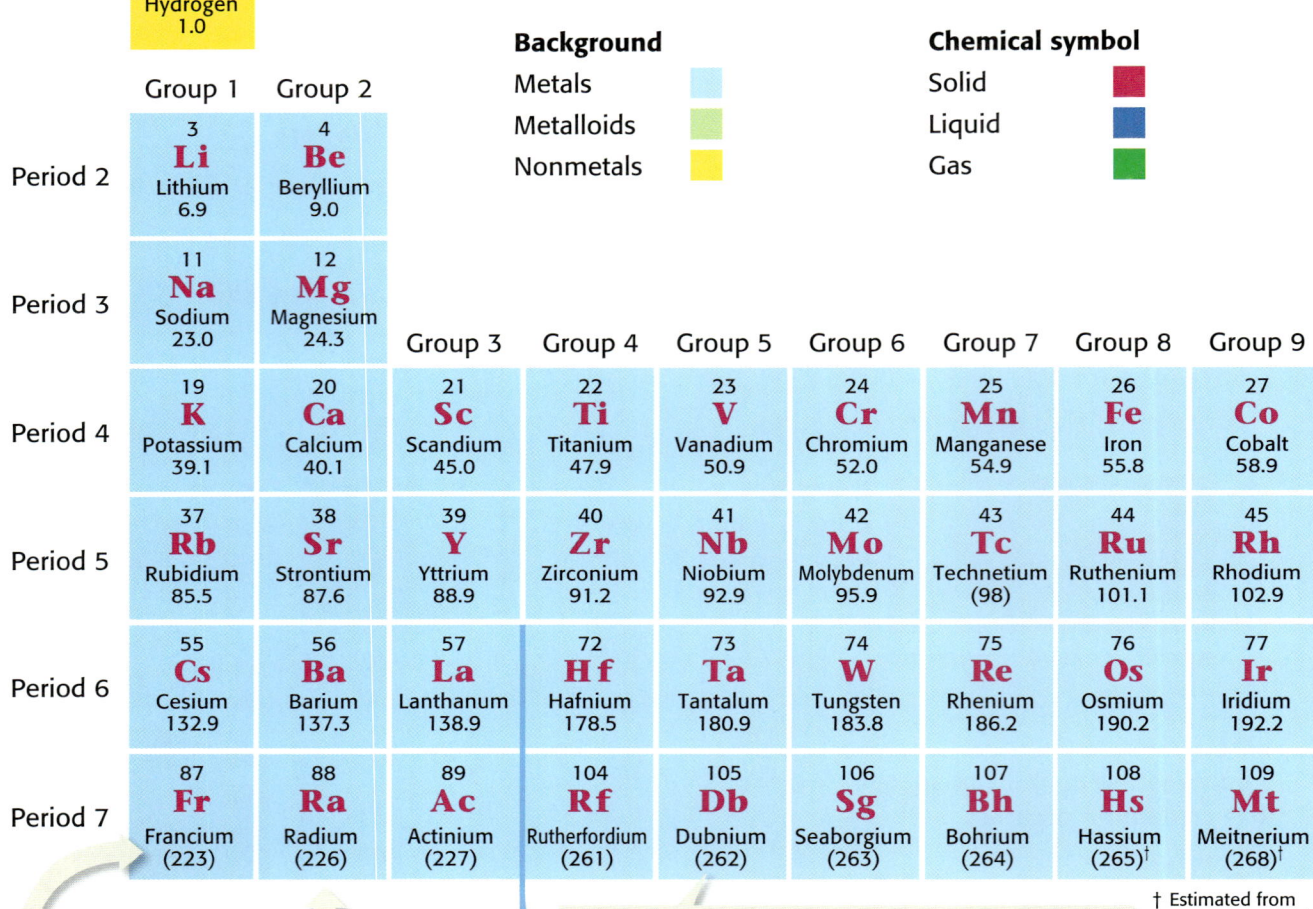

Period 1	1 **H** Hydrogen 1.0

	Group 1	Group 2
Period 2	3 **Li** Lithium 6.9	4 **Be** Beryllium 9.0
Period 3	11 **Na** Sodium 23.0	12 **Mg** Magnesium 24.3

	Group 3	Group 4	Group 5	Group 6	Group 7	Group 8	Group 9
Period 4 (K 19, Ca 20)	21 **Sc** Scandium 45.0	22 **Ti** Titanium 47.9	23 **V** Vanadium 50.9	24 **Cr** Chromium 52.0	25 **Mn** Manganese 54.9	26 **Fe** Iron 55.8	27 **Co** Cobalt 58.9
Period 5 (Rb 37, Sr 38)	39 **Y** Yttrium 88.9	40 **Zr** Zirconium 91.2	41 **Nb** Niobium 92.9	42 **Mo** Molybdenum 95.9	43 **Tc** Technetium (98)	44 **Ru** Ruthenium 101.1	45 **Rh** Rhodium 102.9
Period 6 (Cs 55, Ba 56)	57 **La** Lanthanum 138.9	72 **Hf** Hafnium 178.5	73 **Ta** Tantalum 180.9	74 **W** Tungsten 183.8	75 **Re** Rhenium 186.2	76 **Os** Osmium 190.2	77 **Ir** Iridium 192.2
Period 7 (Fr 87, Ra 88)	89 **Ac** Actinium (227)	104 **Rf** Rutherfordium (261)	105 **Db** Dubnium (262)	106 **Sg** Seaborgium (263)	107 **Bh** Bohrium (264)	108 **Hs** Hassium (265)†	109 **Mt** Meitnerium (268)†

Period 4: 19 **K** Potassium 39.1, 20 **Ca** Calcium 40.1
Period 5: 37 **Rb** Rubidium 85.5, 38 **Sr** Strontium 87.6
Period 6: 55 **Cs** Cesium 132.9, 56 **Ba** Barium 137.3
Period 7: 87 **Fr** Francium (223), 88 **Ra** Radium (226)

Lanthanides

58 **Ce** Cerium 140.1	59 **Pr** Praseodymium 140.9	60 **Nd** Neodymium 144.2	61 **Pm** Promethium (145)	62 **Sm** Samarium 150.4

Actinides

90 **Th** Thorium 232.0	91 **Pa** Protactinium 231.0	92 **U** Uranium 238.0	93 **Np** Neptunium (237)	94 **Pu** Plutonium (244)

Appendix

Topic: **Periodic Table**
Go To: **go.hrw.com**
Keyword: **HN0 PERIODIC**
Visit the HRW Web site for updates on the periodic table.

Group 18

| 2 **He** Helium 4.0 |

This zigzag line reminds you where the metals, nonmetals, and metalloids are.

	Group 13	Group 14	Group 15	Group 16	Group 17	
	5 **B** Boron 10.8	6 **C** Carbon 12.0	7 **N** Nitrogen 14.0	8 **O** Oxygen 16.0	9 **F** Fluorine 19.0	10 **Ne** Neon 20.2
	13 **Al** Aluminum 27.0	14 **Si** Silicon 28.1	15 **P** Phosphorus 31.0	16 **S** Sulfur 32.1	17 **Cl** Chlorine 35.5	18 **Ar** Argon 39.9

Group 10	Group 11	Group 12						
28 **Ni** Nickel 58.7	29 **Cu** Copper 63.5	30 **Zn** Zinc 65.4	31 **Ga** Gallium 69.7	32 **Ge** Germanium 72.6	33 **As** Arsenic 74.9	34 **Se** Selenium 79.0	35 **Br** Bromine 79.9	36 **Kr** Krypton 83.8
46 **Pd** Palladium 106.4	47 **Ag** Silver 107.9	48 **Cd** Cadmium 112.4	49 **In** Indium 114.8	50 **Sn** Tin 118.7	51 **Sb** Antimony 121.8	52 **Te** Tellurium 127.6	53 **I** Iodine 126.9	54 **Xe** Xenon 131.3
78 **Pt** Platinum 195.1	79 **Au** Gold 197.0	80 **Hg** Mercury 200.6	81 **Tl** Thallium 204.4	82 **Pb** Lead 207.2	83 **Bi** Bismuth 209.0	84 **Po** Polonium (209)	85 **At** Astatine (210)	86 **Rn** Radon (222)
110 **Ds** Darmstadtium (269)†	111 **Uuu** Unununium (272)†	112 **Uub** Ununbium (277)†	114 **Uuq** Ununquadium (285)†					

The names and three-letter symbols of elements are temporary. They are based on the atomic numbers of the elements. Official names and symbols will be approved by an international committee of scientists.

63 **Eu** Europium 152.0	64 **Gd** Gadolinium 157.2	65 **Tb** Terbium 158.9	66 **Dy** Dysprosium 162.5	67 **Ho** Holmium 164.9	68 **Er** Erbium 167.3	69 **Tm** Thulium 168.9	70 **Yb** Ytterbium 173.0	71 **Lu** Lutetium 175.0
95 **Am** Americium (243)	96 **Cm** Curium (247)	97 **Bk** Berkelium (247)	98 **Cf** Californium (251)	99 **Es** Einsteinium (252)	100 **Fm** Fermium (257)	101 **Md** Mendelevium (258)	102 **No** Nobelium (259)	103 **Lr** Lawrencium (262)

Appendix

Glossary

A

abyssal plain a large, flat, almost level area of the deep-ocean basin (48)

alluvial fan a fan-shaped mass of material deposited by a stream when the slope of the land decreases sharply (14)

aquifer a body of rock or sediment that stores groundwater and allows the flow of groundwater (17)

artesian spring a spring whose water flows from a crack in the cap rock over the aquifer (19)

B

benthic environment the region near the bottom of a pond, lake, or ocean (53)

benthos the organisms that live at the bottom of the sea or ocean (52)

C

channel the path that a stream follows (7)

continental rise the gently sloping section of the continental margin located between the continental slope and the abyssal plain (48)

continental shelf the gently sloping section of the continental margin located between the shoreline and the continental slope (48)

continental slope the steeply inclined section of the continental margin located between the continental rise and the continental shelf (48)

Coriolis effect the apparent curving of the path of a moving object from an otherwise straight path due to the Earth's rotation (82)

D

deep current a streamlike movement of ocean water far below the surface (83)

delta a fan-shaped mass of material deposited at the mouth of a stream (13)

deposition the process in which material is laid down (12)

desalination (DEE SAL uh NAY shuhn) a process of removing salt from ocean water (61)

divide the boundary between drainage areas that have streams that flow in opposite directions (6)

E

El Niño a change in the surface water temperature in the Pacific Ocean that produces a warm current (88)

erosion the process by which wind, water, ice, or gravity transports soil and sediment from one location to another (4)

F

floodplain an area along a river that forms from sediments deposited when the river overflows its banks (14)

L

La Niña a change in the eastern Pacific Ocean in which the surface water temperature becomes unusually cool (88)

load the materials carried by a stream; *also* the mass of rock overlying a geological structure (8)

longshore current a water current that travels near and parallel to the shoreline (93)

M

mid-ocean ridge a long, undersea mountain chain that forms along the floor of the major oceans (49)

N

neap tide a tide of minimum range that occurs during the first and third quarters of the moon (98)

nekton all organisms that swim actively in open water, independent of currents (52)

nonpoint-source pollution pollution that comes from many sources rather than from a single, specific site (22, 64)

O

ocean current a movement of ocean water that follows a regular pattern (80)

ocean trench a steep and long depression in the deep-sea floor that runs parallel to a chain of volcanic islands or a continental margin (49)

P

pelagic environment in the ocean, the zone near the surface or at middle depths, beyond the sublittoral zone and above the abyssal zone (56)

permeability the ability of a rock or sediment to let fluids pass through its open spaces, or pores (17)

plankton the mass of mostly microscopic organisms that float or drift freely in freshwater and marine environments (52)

point-source pollution pollution that comes from a specific site (22, 65)

porosity the percentage of the total volume of a rock or sediment that consists of open spaces (17)

R

recharge zone an area in which water travels downward to become part of an aquifer (18)

rift valley a long, narrow valley that forms as tectonic plates separate (49)

S

salinity a measure of the amount of dissolved salts in a given amount of liquid (40)

seamount a submerged mountain on the ocean floor that is at least 1,000 m high and that has a volcanic origin (49)

septic tank a tank that separates solid waste from liquids and that has bacteria that break down the solid waste (25)

sewage treatment plant a facility that cleans the waste materials found in water that comes from sewers or drains (24)

spring tide a tide of increased range that occurs two times a month, at the new and full moons (98)

storm surge a local rise in sea level near the shore that is caused by strong winds from a storm, such as those from a hurricane (95)

surface current a horizontal movement of ocean water that is caused by wind and that occurs at or near the ocean's surface (81)

swell one of a group of long ocean waves that have steadily traveled a great distance from their point of generation (94)

T

tidal range the difference in levels of ocean water at high tide and low tide (98)

tide the periodic rise and fall of the water level in the oceans and other large bodies of water (96)

tributary a stream that flows into a lake or into a larger stream (6)

tsunami a giant ocean wave that forms after a volcanic eruption, submarine earthquake, or landslide (94)

U

undertow a subsurface current that is near shore and that pulls objects out to sea (93)

upwelling the movement of deep, cold, and nutrient-rich water to the surface (87)

W

water cycle the continuous movement of water from the ocean to the atmosphere to the land and back to the ocean (5, 43)

watershed the area of land that is drained by a water system (6)

water table the upper surface of underground water; the upper boundary of the zone of saturation (16)

whitecap the bubbles in the crest of a breaking wave (94)

Spanish Glossary

A

abyssal plain/llanura abisal un área amplia, llana y casi plana de la cuenca oceánica profunda (48)

alluvial fan/abanico aluvial masa de materiales rocosos en forma de abanico, depositados por un arroyo cuando la pendiente del terreno disminuye bruscamente (14)

aquifer/acuífero un cuerpo rocoso o sedimento que almacena agua subterránea y permite que fluya (17)

artesian spring/manantial artesiano un manantial en el que el agua fluye a partir de una grieta en la capa de rocas que se encuentra sobre el acuífero (19)

B

benthic environment/ambiente béntico la región que se encuentra cerca del fondo de una laguna, lago u océano (53)

benthos/benthos los organismos que viven en el fondo del mar o del océano (52)

C

channel/canal el camino que sigue un arroyo (7)

continental rise/elevación continental la sección del margen continental que tiene un ligero declive, ubicada entre el talud continental y la llanura abisal (48)

continental shelf/plataforma continental la sección del margen continental que tiene un ligero declive, ubicada entre la costa y el talud continental (48)

continental slope/talud continental la sección del margen continental que tiene una gran inclinación, ubicada entre la elevación continental y la plataforma continental (48)

Coriolis effect/efecto de Coriolis la desviación aparente de la trayectoria recta que experimentan los objetos en movimiento debido a la rotación de la Tierra (82)

D

deep current/corriente profunda un movimiento del agua del océano que es similar a una corriente y ocurre debajo de la superficie (83)

delta/delta un depósito de materiales rocosos en forma de abanico ubicado en la desembocadura de un río (13)

deposition/deposición el proceso por medio del cual un material se deposita (12)

desalination/desalación (o desalinización) un proceso de remoción de sal del agua del océano (61)

divide/división el límite entre áreas de drenaje que tienen corrientes que fluyen en direcciones opuestas (6)

E

El Niño/El Niño un cambio en la temperatura del agua superficial del océano Pacífico que produce una corriente caliente (88)

erosion/erosión el proceso por medio del cual el viento, el agua, el hielo o la gravedad transporta tierra y sedimentos de un lugar a otro (4)

F

floodplain/llanura de inundación un área a lo largo de un río formada por sedimentos que se depositan cuando el río se desborda (14)

L

La Niña/La Niña un cambio en el océano Pacífico oriental por el cual el agua superficial se vuelve más fría que de costumbre (88)

load/carga los materiales que lleva un arroyo; también, la masa de rocas que recubre una estructura geológica (8)

longshore current/corriente de ribera una corriente de agua que se desplaza cerca de la costa y paralela a ella (93)

M

mid-ocean ridge/dorsal oceánica una larga cadena submarina de montañas que se forma en el suelo de los principales océanos (49)

N

neap tide/marea muerta una marea que tiene un rango mínimo, la cual ocurre durante el primer y el tercer cuartos de la Luna (98)

nekton/necton todos los organismos que nadan activamente en las aguas abiertas, de manera independiente de las corrientes (52)

nonpoint-source pollution/contaminación no puntual contaminación que proviene de muchas fuentes, en lugar de provenir de un solo sitio específico (22, 64)

O

ocean current/corriente oceánica un movimiento del agua del océano que sigue un patrón regular (80)

ocean trench/fosa oceánica una depresión empinada y larga del suelo marino profundo, paralela a una cadena de islas volcánicas o al margen continental (49)

P

pelagic environment/ambiente pelágico en el océano, la zona ubicada cerca de la superficie o en profundidades medias, más allá de la zona sublitoral y por encima de la zona abisal (56)

permeability/permeabilidad la capacidad de una roca o sedimento de permitir que los fluidos pasen a través de sus espacios abiertos o poros (17)

plankton/plancton la masa de organismos en su mayoría microscópicos que flotan o se encuentran a la deriva en ambientes de agua dulce o marina (52)

point-source pollution/contaminación puntual contaminación que proviene de un lugar específico (22, 65)

porosity/porosidad el porcentaje del volumen total de una roca o sedimento que está formado por espacios abiertos (17)

R

recharge zone/zona de recarga un área en la que el agua se desplaza hacia abajo para convertirse en parte de un acuífero (18)

rift valley/fosa tectónica un valle largo y estrecho que se forma cuando se separan las placas tectónicas (49)

S

salinity/salinidad una medida de la cantidad de sales disueltas en una cantidad determinada de líquido (40)

seamount/montaña submarina una montaña sumergida que se encuentra en el fondo del océano, la cual tiene por lo menos 1,000 m de altura y cuyo origen es volcánico (49)

septic tank/tanque séptico un tanque que separa los desechos sólidos de los líquidos y que tiene bacterias que descomponen los desechos sólidos (25)

sewage treatment plant/planta de tratamiento de residuos una instalación que limpia los materiales de desecho que se encuentran en el agua procedente de cloacas o alcantarillas (24)

spring tide/marea muerta una marea de mayor rango que ocurre dos veces al mes, durante la luna nueva y la luna llena (98)

storm surge/marea de tempestad un levantamiento local del nivel del mar cerca de la costa, el cual es resultado de los fuertes vientos de una tormenta, como por ejemplo, los vientos de un huracán (95)

surface current/corriente superficial un movimiento horizontal del agua del océano que es producido por el viento y que ocurre en la superficie del océano o cerca de ella (81)

swell/mar de leva un grupo de olas oceánicas grandes que se han desplazado una gran distancia desde el punto en el que se originaron (94)

T

tidal range/rango de marea la diferencia en los niveles del agua del océano entre la marea alta y la marea baja (98)

tide/marea el ascenso y descenso periódico del nivel del agua en los océanos y otras masas grandes de agua (96)

tributary/afluente un arroyo que fluye a un lago o a otro arroyo más grande (6)

tsunami/tsunami una ola gigante del océano que se forma después de una erupción volcánica, terremoto submarino o desprendimiento de tierras (94)

U

undertow/resaca un corriente subsuperficial que está cerca de la orilla y que arrastra los objetos hacia el mar (93)

upwelling/surgencia el movimiento de las aguas profundas, frías y ricas en nutrientes hacia la superficie (87)

water cycle/ciclo del agua el movimiento continuo del agua: del océano a la atmósfera, de la atmósfera a la tierra y de la tierra al océano (5, 43)

watershed/cuenca hidrográfica el área del terreno que es drenada por un sistema de agua (6)

water table/capa freática el nivel más alto del agua subterránea; el límite superior de la zona de saturación (16)

whitecap/cabrillas las burbujas de la cresta de una ola rompiente (94)

Spanish Glossary

Index

Boldface page numbers refer to illustrative material, such as figures, tables, margin elements, photographs, and illustrations.

Index

Index

W

warm-water currents, 83, **83, 85,** 86, **86**
water. *See also* groundwater; ocean water
  agricultural use of, 26, **26**
  boiling point of, 125
  conservation of, **18,** 27
  drinkable, 22
  filters, 35
  freezing point of, 125
  fresh, 4–27
  household use of, 22, 25, **25,** 27
  industrial use of, 26
  labs on, **17, 23,** 28–29
  pollution, 22–23, **22, 23,** 35
  quality, 23
  river deposition and, 12–15, **12, 13, 14**
  river stages and, 9–10, **9, 10**
  river systems and, 6, **6**
  treatment, 24–25, **24, 25**
  underground, 16–21
  usage, 22–27
  water cycle, 5, **5,** 28–29, 43, **43**
water conservation, **18,** 27
water cycle, 5, **5,** 28–29, 43, **43**
water filters, 35
watersheds, 6, **6**
water table, 16, **16,** 18, **19**
wave energy, 63
wave height, 90, **90**

wavelengths, 90–92, **90, 91, 92**
wave periods, 91, **91**
waves, ocean, 90–95
  breakers, 92, **92**
  deep-water and shallow-water, 92, **92**
  energy of, 63
  formation and movement of, 91, **91**
  lab on, **93**
  longshore currents, 93, **93**
  open-ocean, 94, **94**
  parts of, 90, **90**
  shore currents, 93, **93**
  speed of, 91, **91**
  storm surges, 95, **95**
wave speed, 91, **91**
wave troughs, 90, **90**
weather, condensation and, **5.** *See also* climate
wells, 19, **19**
whales, 107
whitecaps, 94, **94**
winds, 81–82, **81, 82**

Y

Yellowstone National Park, **9**

Z

zone of aeration, 16, **16**
zone of saturation, 16, **16**
zooplankton, 52, **52**

Credits

Abbreviations used: (t) top, (c) center, (b) bottom, (l) left, (r) right, (bkgd) background

PHOTOGRAPHY

Front Cover Ralph A. Clevenger/Corbis

Skills Practice Lab Teens Sam Dudgeon/HRW

Connection to Astrology Corbis Images; **Connection to Biology** David M. Phillips/Visuals Unlimited; **Connection to Chemistry** Digital Image copyright © 2005 PhotoDisc; **Connection to Environment** Digital Image copyright © 2005 PhotoDisc; **Connection to Geology** Letraset Phototone; **Connection to Language** Arts Digital Image copyright © 2005 PhotoDisc; **Connection to Meteorology** Digital Image copyright © 2005 PhotoDisc; **Connection to Oceanography** © ICONOTEC; **Connection to Physics** Digital Image copyright © 2005 PhotoDisc

Table of Contents iv (cl), Glenn M. Oliver/Visuals Unlimited; iv (b), Stuart Westmorland/CORBIS; v (t), Darrell Wong/Getty Images/Stone; x (bl), Sam Dudgeon/HRW; xi (tl), John Langford/HRW; xi (b), Sam Dudgeon/HRW; xii (tl), Victoria Smith/HRW; xii (bl), Stephanie Morris/HRW; xii (br), Sam Dudgeon/HRW; xiii (tl), Patti Murray/Animals, Animals; xiii (tr), Jana Birchum/HRW; xiii (b), Peter Van Steen/HRW

Chapter One 2-3, Owen Franklin/CORBIS; 4, Tom Bean/DRK Photo; 6, E.R.I.M./Stone; 7, Jim Wark/Peter Arnold; 7 (tl), Nancy Simmerman/Getty Images/Stone; 9, Frans Lanting/Minden Pictures; 9 (cr), Laurence Parent; 10 (t), The G.R. "Dick" Roberts Photo Library; 10, Galen Rowell/Peter Arnold, Inc.; 11 (t), Nancy Simmerman/Getty Images/Stone; 12, Glenn M. Oliver/Visuals Unlimited; 13 (t), The Huntington Library/SuperStock; 13 (b), Earth Satellite Corporation/Science Photo Library/Photo Researchers, Inc.; 14 (t), Visuals Unlimited/Martin G. Miller; 14 (b), Earth Satellite Corporation; 15, Jerry Laizure/AP/Wide World Photos; 20, Rich Reid/Animals Animals/Earth Scenes; 21, Leif Skoogfers/Woodfin Camp & Associates, Inc.; 22, Digital Image © 2005, Eyewire/Getty Images; 23, Morton Beebe/CORBIS; 27, Getty Images/Stone; 28, Victoria Smith/HRW; 30, Martin Harvey; Gallo Images/CORBIS; 31 (t), The Huntington Library/SuperStock; 31 (b), Jim Wark/Peter Arnold; 34 (t), David R. Parks; 34 (br), Martin Harvey; Gallo Images/CORBIS; 35 (tr), Photo by Sam Kittner, courtesy of Rita Colwell/National Science Foundation; 35 (b), Anwar Huq, UMBI

Chapter Two 36-37, Henry Wolcott/Getty Images/National Geographic; 38, Tom Van Sant, Geosphere Project/Planetary Visions/Science Photo Library; 42 (l), U.S. Navy; 42(r), U.S. Navy; 44, Rosentiel School of Marine and Atmospheric Science, University of Miami; 45, Courtesy of Robert Cantor/Christian Grantham; 46 (l), W. Haxby, Lamont-Doherty Earth Observatory/ Science Photo Library/Photo Researchers, Inc.; 47 (t), NOAA/NSDS; 50 (r), James Wilson/Woodfin Camp & Associates; 50 (l), Norbert Wu; 53, Stuart Westmorland/CORBIS; 53, Stuart Westmorland/CORBIS; 54 (t), Mike Bacon/Tom Stack & Associates; 54 (b), James B. Wood; 55 (b), JAMESTEC; 56, Mike Hill/Getty Images/Photographer's Choice; 57, ©2005 Norbert Wu/www.norbertwu.com; 58, Joel W. Rogers; 59 (t), Breg Vaughn/Tom Stack & Associates; 59 (b), Gregory Ochocki/Photo Researchers, Inc.; 60, Terry Vine/Getty Images/Stone; 61, Steve Raymer/National Geographic Society Image Collection; 62 (tl), Institute of Oceanographic Sciences/NERC/Science Photo Library/Photo Researchers, Inc.; 62 (inset), Charles D. Winters/Photo Researchers, Inc.; 63, Gregory Ochocki/Photo Researchers, Inc.; 64 (r), Photo Edit; 64 (l), Andy Christiansen/HRW; 64 (c), Richard Hamilton Smith/CORBIS; 65 (t), E. R. Degginger/Color-Pic, Inc.; 65 (b), Fred Bavendam/Peter Arnold, Inc.; 66, Greenpeace International; 67 (t), Ben Osborne/Getty Images/Stone; 67 (b), Courtesy Mobil; 68 (l), Courtesy Texas General Land Office Adopt-A-Beach Program; 68 (r), Tony Amos; 69, Ben Osborne/Getty Images/Stone; 71, Sam Dudgeon/HRW; 72, ©2005 Norbert Wu/www.norbertwu.com; 76 (b), © Reuters NewMedia Inc./CORBIS; 76 (tr), ©Patricia Jordan/Peter Arnold, Inc.; 76 (tl), ©Aldo Brando/Peter Arnold, Inc.; 77 (r), HO/The Cousteau Society/Reuters Photo Archive/NewsCom; 77 (l), Parrot Pascal/Corbis Sygma

Chapter Three 78-79, Tom Salyer/Reuters NewMedia Inc./CORBIS; 80, Hulton Archive/Getty Images; 81 (r), Sam Dudgeon/HRW; 81 (t), Rosentiel School of Marine and Atmospheric Science, University of Miami; 88, Lacy Atkins/San Francisco Examiner/AP/Wide World Photos; 93 (b), CC Lockwood/Bruce Coleman, Inc.; 94 (tl), Darrell Wong/Getty Images/Stone; 94 (tr), August Upitis/Getty Images/Taxi; 99 (tl), VOSCAR/The Maine Photographer; 99 (tr), VOSCAR/The Maine Photographer; 100, Andy Christiansen/HRW; 106 (t), J.A.L. Cooke/Oxford Scientific Films/Animals Animals/Earth Scenes; 107 (t), Pacific Whale Foundation; 107 (b), Flip Nicklin/Minden Pictures

Lab Book/Appendix 110, Victoria Smith/HRW; 111, Andy Christiansen/HRW; 112, Sam Dudgeon/HRW; 115, Victoria Smith/HRW

TEACHER EDITION CREDITS

1E (l), Tom Bean/DRK Photo; 1E (r), Earth Satellite Corporation/Science Photo Library/Photo Researchers, Inc.; 1F (r), Getty Images/Stone; 35E (l), Tom Van Sant, Geosphere Project/Planetary Visions/Science Photo Library; 35E (br), James Wilson/Woodfin Camp & Associates; 35E (bl), Norbert Wu; 35F (l), Mike Hill/Getty Images/Photographer's Choice; 35F (r), Ben Osborne/Getty Images/Stone; 77E (bl), Hulton Archive/Getty Images; 77E (r), Sam Dudgeon/HRW; 77F (tl), Darrell Wong/Getty Images/Stone; 77F (tr), August Upitis/Getty Images/Taxi; 77F (br), VOSCAR/The Maine Photographer; 77F (br), VOSCAR/The Maine Photographer

Answers to Concept Mapping Questions

The following pages contain sample answers to all of the concept mapping questions that appear in the Chapter Reviews. Because there is more than one way to do a concept map, your students' answers may vary.

CHAPTER 1   **The Flow of Fresh Water**

16.

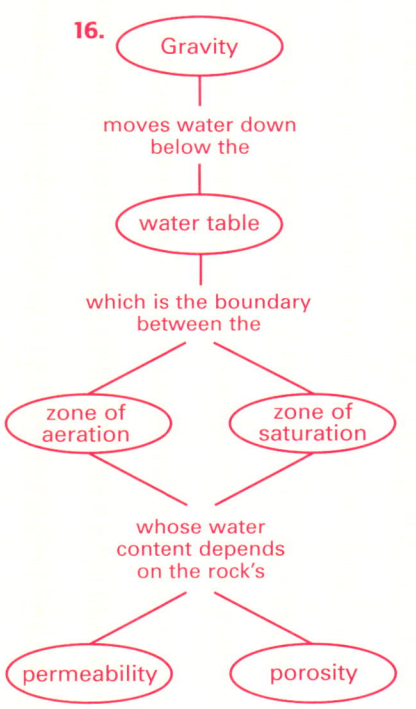

CHAPTER 2   **Exploring the Oceans**

19.

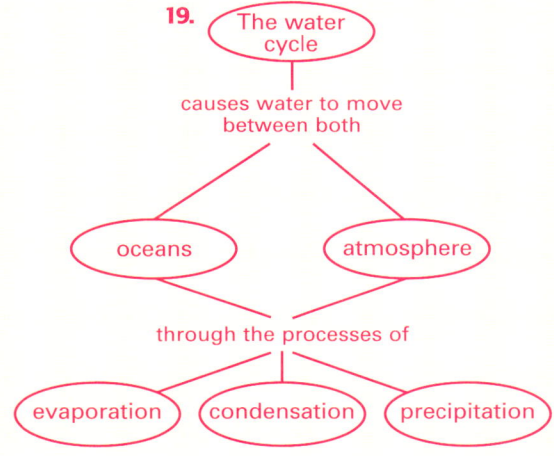

CHAPTER 3   **The Movement of Ocean Water**

17.

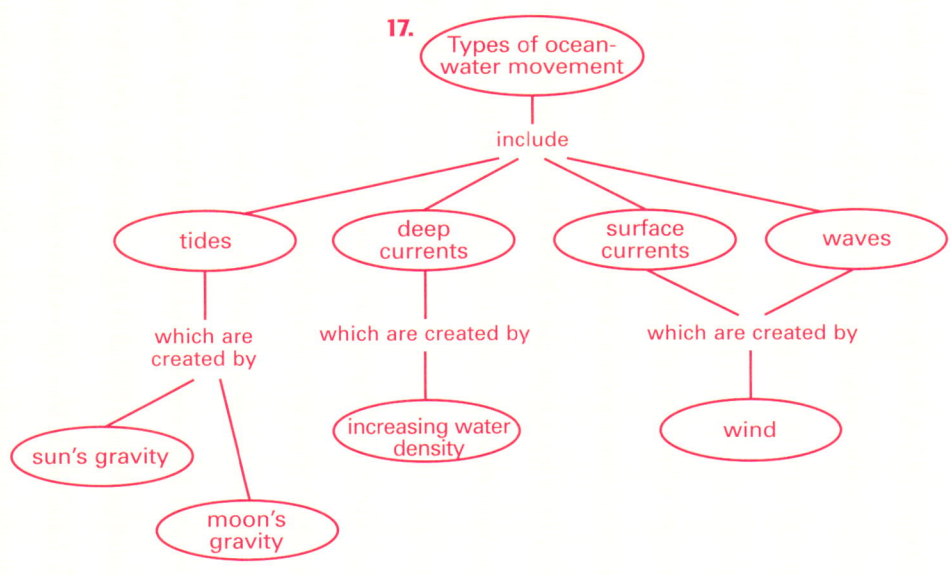